A Life in Science

The author, about 1960

A Life in Science

Sir Nevill Mott

Taylor & Francis
London and Philadelphia
1986

UK   Taylor & Francis Ltd, 4 John St, London WC1N 2ET

USA   Taylor & Francis Inc., 242 Cherry St, Philadelphia, PA 19106-1906

Copyright © 1986 N. F. Mott

Reprinted 1987

British Library Cataloguing in Publication Data

Mott, N.F.
   A life in science.
   1. Mott, N.F.   2. Physicists—England—
   Biography
   I. Title
   530′.092′4   QC16.M/

   ISBN 0-85066-333-4

Library of Congress Cataloging in Publication Data is available

Typeset by Mathematical Composition Setters Ltd, Salisbury, Wilts
Printed in Great Britain by Taylor & Francis (Printers) Ltd, Basingstoke,
Hants.

Contents

Contents

Preface

Why did I write this book? It is because I have lived through a heroic period in the development of science, and was associated particularly with the beginnings of solid state physics. The most dramatic story in twentieth century physics is that of Rutherford's discovery of the nucleus, leading to the discovery of the neutron, nuclear fission and the demonstration by Frisch and Peierls that a uranium bomb could be made. The effects on warfare have been potentially enormous. The story has often been told and biographies of many of the chief protagonists abound. Less has been written about the beginnings of solid state physics. None the less it could be said that microelectronics, based on solid state physics, has affected our civilisation hardly less profoundly. Telecommunications, the computer, automation all depend on it, as does advanced military technology. Many benefits have come from all this and some damage, including an increase in unemployment, one hopes temporary, where electronics has taken over from men and women.

Those who put together the ideas on which all this was based could foresee little or nothing of the consequences, particularly up till the beginning of the war of 1939–45. Our reasons for choosing to study this subject will be a major theme of this book, since I believe that my own experience throws light on it. After the war we knew only too well that what we had done was of industrial and military importance and we saw, mainly in the USA, the development of the transistor and all that followed from it. I shall explain how this affected my attitude to collaboration with industry.

Since this is an autobiography, I shall describe how a life in science affected my attitude to other matters. Our Cambridge colleges are dedicated to Education, Religion, Learning and Research, and since all have been my concern, I have written about them. I have tried to write as far as possible without the use of technical language, especially when describing research, in the hope that 60 years in physics may be of interest to some readers outside this particular discipline.

Many members of my family and friends have read all or parts of this book in manuscript and have made very helpful suggestions. I am particularly indebted to Margaret Gowing, Professor of the History of Science at Oxford University, who encouraged me to write the book in the first place and who has throughout given most valuable help and advice.

CHAPTER ONE

Parentage and childhood

Ever since I knew what physics was, I have wanted to be a physicist. In this I was influenced by my parents, much more than by any teacher at school. My father and my mother both worked in the Cavendish Laboratory at Cambridge under J. J. Thomson, and, though neither made a career in science, they were both influenced by this experience and kept in touch with what was going on. Quite early in my life they communicated to me the excitement and importance of the subject.

My father was the eldest of a family of three; neither his parents nor his sister and brother went to a university. He was educated at Reigate Grammar School and Kings College, London, and from there he got a scholarship to Trinity College, Cambridge, to work in the Cavendish. In 1902, only a few years earlier, Thomson had discovered the electron, and he suggested to my father an experiment to determine how many of these particles there were in a metal. Though some sixty years later experiments of this kind have been performed successfully (see Chapter 16), they were then beyond the capacity of physics equipment, and my father got no result. No job in Cambridge was available, and my father, who married at that time (1904), accepted the post of senior science master at Giggleswick, a small independent school in Yorkshire. He moved from Giggleswick seven years later and went into educational administration, first in 1911 in Stafford, and then as director of education for Cheshire in 1918 and for the city of Liverpool in 1922, from where he retired in 1945.

Concerning the origins of our family, my father believed that the Motts were of French Huguenot descent and that they had been expelled from France after the revocation of the Edict of Nantes. Although this appears to be believed by other Motts, not closely related, we have no documentary proof of it. My father made efforts to trace our ancestry in his old age, but we have not been able to get beyond about 1700. Perhaps Isaac Mott, maker of pianos in Brighton in the reign of George IV, was our most distinguished ancestor. About 1982, I received a letter out of the blue from a Compte de la Motte, living at Nice, asking if we were related and offering to exchange information about our ancestors; but his family, going back to the crusades, were all loyal Catholic subjects of the King of France and there was no sign of any Huguenot.

My mother, Lilian Mary Reynolds, was born and brought up in the small Norfolk town of Thetford, the youngest of a family of six. None were scientists, one was a successful business man. Her most distinguished ancestor was her grandfather, John Richardson (1787–1865), Fellow of the Royal Society and Arctic explorer. A brilliant girl, she was educated at Cheltenham Ladies' College, the first girl from Thetford Grammar School, she believed, to go away to school. From there she went to Newnham College, Cambridge, to read mathematics, where she was coached by the famous Philippa Fawcett, who had been 'above Senior Wrangler' in the mathematical tripos (the Cambridge honours degree examination) some years previously. In those days candidates were listed in the order of their marks, but women were not full members of the University and so were placed in the tripos list as equal to one of the men. My mother was equal to eleventh, but was rather disappointed by that. Then she tried her hand at physics, and the photographs of J. J. Thomson's research groups for 1903 and 1904, which hang in the present Cavendish Laboratory (see Figure 2), show

Figure 1. The author as a small child with his mother and father.

P.S.Barlow K.Przibram. E.P.Adams. F.Horton. B.Davis. W.Makower. A.Wood. R.Hosking.

J.Blyth. J.A.Cunningham. R.K.McClung. J.J.E.Durack. O.W.Richardson. C.F.Mott. G.W.Walker. W.H.White. N.R.Campbell. W.M.Varley.

J.B.B.Burke. H.A.Wilson. Miss Brooks. Prof.J.J.Thomson. Miss Reynolds. P.V.Bevan. J.H.Vincent.

S.C.Laws. J.Strachan. G.Jaffé. G.Owen.

Figure 2. Cavendish Research group, showing the author's father (C.F. Mott) and mother (Miss Reynolds).

both her and my father. He had coached her in physics; they became engaged and, after she had taught for a year at St. Paul's Girls' school in London, they married in 1904, and started married life together in Giggleswick. I was born on 30th September 1905 in a Leeds nursing home and my sister Joan was born in Giggleswick two years later. We lived in a large cottage, across the road from the shallow 'beck' into which my sister and I used to fall from time to time. Our house, as I remember, had no bathroom, but on the stipend of a science master we could afford a nurse and a maid. Starting at this time my mother was throughout her life increasingly involved in social causes, such as votes for women (until it was achieved), child adoption, family planning at a time when this was widely opposed, the plight of refugees during the Nazi period and many others. She and my father were, it seems to me, people of the highest moral standards; but, after a wedding in a Unitarian church with a service they wrote themselves, they seem to have abandoned any form of religious belief that needed expression in church. My mother's early letter to a friend, who was a missionary in India, reveal her reading books such as Albert Revilles *Jésus de Nazareth*, which led her to a way of thinking in which Jesus was a great teacher but not, in any sense that she could understand, the Son of God. They show too her fear that her beliefs would give pain to her parents and to her brothers, one of whom was ordained. I do not remember ever going to church with my parents, and I was not baptized or confirmed until after my retirement.

My mother died on 11 November 1952, in a Bath nursing home at the age of 73. My father, who lived fifteen years longer, wrote a memoir, privately printed, recording her early life, her religious opinions when at college, and her social work. He also endowed an L. M. Mott prize at Newnham College for the student with the best performance in physics each year. A copy of the memoir is also presented to each prize-winner.

From my childhood in Giggleswick, my earliest memories are of the Pennine hills, green slopes and stone walls. All through my schooldays our August family holiday was normally spent somewhere in the hills, in North Wales, Derbyshire or the Lake District, and in the school holidays I could usually get to the hills alone with my bicycle. The roads were still fairly free of cars then. Throughout my life green hills and stone walls, with small streams and waterfalls, have seemed the most welcome face of England.

CHAPTER TWO

School

When I was six my parents moved to Stafford, but, because of some anxiety about my health, I did not go to school until I was ten; up till then my mother gave me lessons at home. Except for my sister, I do not remember meeting other children at that time. The school, a small preparatory school (Baswich House), was near enough to Stafford for me to be a weekly boarder and I was there in the first year of the school's existence. There were only three boys, and we never rose above thirty. The headmaster and owner of the school, G. F. A. Osborn, must have been an excellent teacher, taking me well on in algebra and Latin before I left, and, as far as I can remember, even introducing me to calculus. When I was thirteen I was put in for some scholarship examinations. I failed to get one at Repton, but was successful at Clifton College, where I spent five years.

I remember my childhood as very happy, apart from some years at Clifton. My parents chose it because they had heard that the laboratories were good. However I do not remember finding much inspiration in the physics master and E. J. Holmyard, the chemist, though distinguished, seemed mainly interested in the Arabic origins of his science. However we had a superb mathematics teacher, H. C. Beaven, and it was his influence, I am sure, that turned me towards mathematical physics, and showed me that mathematics was something I could do well. There is not much else that I can remember with gratitude. My parents were persuaded to put me on the classical side till I was 16, because it was thought that I would not get enough competition elsewhere. I therefore left school with some Latin and Greek but not enough to read with pleasure and with only the most rudimentary French. Life outside the classroom in an English public school just after the first world war has often been described. C. S. Lewis's account of Malvern in *Surprised by Joy* is just as I remember it. On the first day all new boys were summoned to meet the head of the house, and it was explained to us what our purpose in life was to be. The school was divided into 'houses', we were in School House, and our life's purpose was to serve the house. We

Figure 3. The author, aged about 8 with his sister Joan (aged 6).

could achieve a place in one of the house teams for Rugby football, and serve the house in that way, or in cricket. If we failed in that there were the less prestigious games, fives or tennis. And right at the bottom, for those who could not achieve any of these things, there was the house music competition. Music, it seemed, had achieved respectability by becoming a game. But work in class had not; it did not 'serve the house'; there was no 'house mathematical competition'. Then there was the almost obligatory romantic attachment of each of the older boys to one of the more good looking younger ones. The question of who was with whom was an all-pervading subject of conversation. Games consisted too often of the compulsory watching of house matches; not to be there was disloyalty to the house. When the weather was bad there were 'runs' across the Clifton Downs; one term I took to cutting them and exploring the fascinating old city of Bristol, destroyed now by the wartime bombing and the subsequent development; but I was eventually caught and punished by the prefects.

But of course there were good bits. General Sir Douglas Haig was our most eminent 'old boy' and the Officers Training Corps was very much in evidence. We were an engineering corps, making bridges across the pond in the Clifton zoo, and that was fun. There was a 'knotting and lashing competition', in which we had to make a 'bowline on a bight' and other knots in (was it?) six seconds. Here was a way to serve the house, at which by dint of hours and hours of practice I became quite proficient. As the years passed friendship and conversation about things other than sport and the rest became possible. I could get away to the school library to read what I liked. I remember meeting there a boy from another house, who confessed that he liked reading poetry—but I was not to tell anyone. Mathematics with Beaven—'Fuzzy B' we called him—was always an excitement, and here the exchange of ideas began to give me the intense pleasure that it has afforded all through my life in research. I was about sixteen and the problems of why $2^{\frac{1}{2}}$ meant the square root of two and why 2^0 was equal to unity were puzzling me. I remember discussing it with another boy in the dormitory at night, when I suddenly saw that $2^{\frac{1}{2}} \times 2^{\frac{1}{2}}$ must be $2^{\frac{1}{2}+\frac{1}{2}}$, shared the idea with him, and felt really excited.

In December 1923 I went to Cambridge to take the scholarship examination in mathematics and physics for entry to St John's College. I enjoyed it very much and was awarded a major scholarship; I learned later that I had the highest marks of that year's candidates. So I left school in that month.

I have never kept a diary, but throughout my life I wrote long letters to my parents, and discovered only when I started this autobiography that they had kept them all, from my last year at school till after my mother's death, when my father came to live with my widowed sister in the same road in Cambridge as myself.

My parents, with their experience both of physics and of public affairs, were intensely interested in all that I did, and I believe very proud of what I achieved. I knew this, and wanted them to know about what I was doing—and indeed to offer me advice. Letters, especially to parents, are perhaps less frank than a diary, as the writer wants to give a good impression, but none the less I shall use them in this book. Those written during this period give a much better impression than my memory of life at school. But I was amused by a letter about the first time I read a lesson in the school chapel, where I say that 'I now sit shaking with anticipation'—but then that I am not really very nervous about it—at least not more so than the boy who was reading the other lesson. Chapel was held twice on Sundays, and gave me a memory of the prayers and hymns of the Church of England, which came back to me when much later in life I started to find church services valuable and meaningful (Chapter 22).

Another letter says that I had been looking at old Cambridge scholarship papers, and found as a general essay subject the quotation

'Ah Meredith, who can define him? His style is chaos illuminated by flashes of lightning. As a writer he has mastered everything except language; as a novelist he can do everything except tell a story; as an artist he is everything except articulate.'

My mother told me that she had called me Nevill after Nevill Beauchamp in Meredith's novel *Beauchamp's Career*. So I must have enjoyed sending her this. I could never get through that novel.

CHAPTER THREE

Cambridge 1924–27

When I left school, after a few months at home reading mathematics with a coach, my parents sent me to Lausanne for the summer to learn French. Nothing could have been more unlike school. I stayed at a small and cheap pension, there were students of varied nationality and both sexes to get to know, mountains to be climbed, the lake (not yet polluted) for boating and swimming. Then, after a month with my parents in the Alps, and another in Liverpool, I went in October to St John's College, Cambridge. Surely no Oxbridge student will ever forget the joy of having a college room, (in fact two rooms), and an 'oak' (the outer door) that could be 'sported' (locked), or opened as one wished. Cambridge was complete and absolute freedom. The rule which seems to have riled some later generations that an undergraduate must be back in college by midnight, was something that would never have occurred to me to question.

As I had a scholarship in mathematics, I read for the mathematical tripos, which was arranged in the following way. Part I, to be taken at the end of the first year, was relatively elementary and scholars were supposed to take it in their stride without going to lectures about the material it contained. They started at once with lectures for Part II. The course was divided between pure and applied mathematics. Pure mathematics was the most exciting. I remember in my first year the thrill of understanding what a limit was, and how in differential calculus it was forbidden to use the word 'small'. This was quite new to me. I was never tempted to become a pure mathematician; physics had always been my aim since hearing about it in childhood, but I knew that the first qualification for theoretical physics was a good understanding of mathematics. In fact, I never learned enough, and missed some skills, such as group theory, which I needed later on. The lectures on applied mathematics, on dynamics and electricity, were less exciting. Part II was usually taken at the end of the third year, qualifying for the degree of B.A. It could be combined with advanced papers on specialized subjects, in a course called Schedule B, (now replaced by Part III of the tripos) for which the award was a B or B*. Together with another student, John Brunyate, I decided to take the whole examination after two years; we were anxious to finish with examinations and begin research, and felt that the risk of not getting a B* was worth it. In the event, we both got a first class honours degree and a B*—but for me I am not sure that it was a good decision. All through my career I have felt that there were bits of mathematics which should have been familiar to me, but which I always postponed mastering.

For us the course consisted of lectures, about two each day, and one hour of supervision each week with a don; two of us went to him together. The quality of supervision was extraordinarily variable, as I believe it often is today. My first supervisor was Harold Jeffreys, later knighted, an exceptionally gifted geophysicist, who became Plumian Professor of Astronomy and Experimental Philosophy in the university, and whose book on *The Earth; its Origin, History and Physical Consitution*, went into six editions. But as a supervisor he was a disappointment. We sat down, and he asked us, 'Have you any questions?'. If we hadn't, he said 'Well; if you haven't any questions perhaps you'd better go.' And usually we did. By far the best supervisor I had was Ebenezer Cunningham, who was the least distinguished in research. I therefore learned of a question which has haunted my whole career: what should be the relation between teaching and research in a university, and is the research man always the best teacher? I remember asking Cunningham what would be the best subjects to choose for Schedule B of the tripos, and how he blew me up; "You're not interested in the Tripos, you're interested in science". So we talked about science, even if I hadn't any questions. I remember, too, a lecture of his in my second year about the electron theory of metals. As in my father's day, we knew that metals contained free electrons and that when a current flows in a wire this is thought to be simply electrons moving along it. But we knew at that time—as they did not in my father's time—that all atoms consist of an atomic nucleus and a certain number of electrons. So insulators, which cannot carry a current, must contain electrons too. In a metal they must be free to move, and in an insulator they must be stuck. I asked Cunningham why this was so—and he told me that it was not understood. It was good to know the limits of knowledge at the time. When just a year or two later 'quantum mechanics' burst upon our particular world, the answer quickly became clear, though a complete understanding of the difference between metals and insulators still eludes us; it has been a subject of my research over half my life.

Cambridge has three terms, ridiculously short (8½ weeks), but scientists were encouraged to come back for five weeks or so of the long vacation. I did this at the end of my first year and I remember it as an idyllic time. Punts on the river, no lectures, interminable discussion with friends about God, the universe, politics, Bernard Shaw, Bertrand Russell and less serious matters. We were supposed to be in by ten o'clock, two hours earlier than in term; curiously, it was eleven for the women's colleges. I remember one night when two of us climbed in after an evening trip on the river—an experience, I suppose, of most undergraduates at that time.

Since, in spite of my mathematics, I had always meant to be a physicist, I started a course of practical physics in the Cavendish, during the vacation, under a famous teacher, Dr G. F. C. Searle, who had known my parents (see fig. 6). He welcomed me with "Behold the second generation"—but I doubt whether he thought much of me; I found the work dull and was confirmed in my ambition to be a theoretical physicist. Searle—like most of us as we get established—was a man of whom stories were told, whether true or not. One story is as follows. A woman student was doing an experiment with a magnetometer and it would not go right. He sniffed around until he guessed the trouble and said, 'You've got iron in your corsets; go and take them off'.

During my second year the General Strike took place. At that time I had a motor bicycle, a second-hand two-cylinder Douglas, which went very well except that the belt from the engine to the back wheel broke from time to time. As I remember, most of us, without thinking deeply on the issues and in spite of vague talk about socialism, were on the side of the government in the strike. I answered a call for volunteers, and I was detailed to patrol the Cambridge–Newmarket road on my motor bicycle. I cannot imagine what I was expected to find. Later we were given tin hats and taken down to Whitechapel, where we slept in a warehouse stacked with lavatory basins, and patrolled the streets. Then the strike ended.

The following letter to my mother describes what I did and, I fear, my rather uninformed feelings about this major event in our social history.

> I'm sorry I didn't write during the strike. I simply didn't think of it. I sent two p.c.s though. One from here about a week ago, and one from Whitechapel, which you don't seem to have got. And a letter last Sunday.
>
> I enclose one or two photographs taken down at Whitechapel—me, and other people, in tin hats and batons. We

have been allowed to keep our batons, memorials of a hearty picnic. I've got mine signed by half our company.

New societies are springing up all over Cambridge. Cambridge docker's club, C.U. conductors. Ours is the C.U. Old Special Constabulary. Motto—Floreat Robertus†; we've got a crest too!

I expect you've heard in the papers about our people all over the place. Of the Dover crowd, mostly boat club, an inspector said 'They drinks like the Devil, they swears like ten devils, and they works like ten thousand devils!' During the strike nobody talked of anything but 'When shall I get away from Cambridge and do something?'

People went and signed at the Guildhall, but they had more people than they wanted there. The thing to do was to form a gang of people who knew each other, with a gang leader, and go to the OTC headquarters to offer one's services. Some engineers got off early. On the second day of the strike, the man who lives opposite me ran into my rooms waving a Union Jack, more or less, with a knapsack of clothes—he was off to work a power station. Unfortunately some hundreds of engineers turned up to work that power station, and he had to come back—the most pathetic incident of the strike! However, he worked his energies off dispatch riding, petrol paid. I did that for two days—that's when I saw Uncle, and also I went to Norwich.

I tried to feel sympathetic to the strikers for two days, and then decided that there really was more to be said for carrying on the country than for not doing so, and also that to stay in Cambridge was dull, and government service was the only means of getting away, the deciding factor, I fear, with most people, low or highbrow. After all, which side can one be enthusiastic for in such a dispute? Thank the Lord for one or two clear cut issues, where there is nothing to be said for the other side,—Birth control v. the Antis. for instance.

The actual picnic in Whitechapel was really great fun, though we had nothing to do. We all went down in private cars; I was in a little Morris Cowley, which stuck on the hills—not that there were any! Rendez-vous outside Lords, where a pilot got on one car to guide us down there.

We inhabited two warehouses, one on each side of a street, and about five stories each. On ours were housed lead piping and fittings, W.C. basins, taps, etc. Amid all this we arranged our palliasses—and slept, more or less. (But when I got home I slept from 11 p.m. to 12.45 p.m.) Programme for the day was bridge,

†Robertus for 'Bobby', the British policeman

football on the roof, having one's photo taken, and queue-ing up for meals, pay, tin hats etc. I was 'on guard' one night, 10.45–12.15 and 3–4.30 a.m. the first guard was amusing—standing keeping back the crowd who would have liked to come up the street and look at us! As for the 3–4.30 one, I didn't appreciate being woken out of my nice blanket by someone's foot, and having to go out and see the East London dawn. I went to sleep very quickly afterwards.

I told you about our march to the Tower for baths, I think. We came home by special train on Saturday. Now I'm doing some work.

When the strike ended, I returned to my work for the tripos, and doubtless thought of little else, though I was never nervous about examinations. But, when I had finished, I knew I had made a bad mistake in one question, and told my parents that I might not get a B*. But I did obtain a first and a B*, to my considerable relief, and turned with a light heart to a summer holiday.

I set off with two friends on my bicycle (a push bicycle this time), going through London to Newhaven, across to Dieppe, and with the aim of getting to Venice. We had a tent on the carrier of my machine. Two of us got there, Frank Oldham (later to be a successful headmaster) and I, while the third fell slightly ill in the Black Forest. We found a doctor for him and, heartlessly, left him behind. Our route was through Rheims, Verdun and some of the still devastated war zone, Metz and Strasbourg, the Arlberg and Brenner passes and the Dolomites. The thousand-mile trip took us three weeks. The villagers were then not used to campers ('We had too much under canvas in the war'), and we could camp on the village green, with the village around us to see us go to bed. In Germany, I remember the disbelief on hearing that we had come all that way on bicycles, '*Immer mit dem Rad?*', they asked.

On the way back we took a train over the Brenner. I had arranged to meet a German student in Munich and go to his house in a village near the Lake of Constance for a month to exchange lessons in our two languages. It turned out that he was the bright son of a peasant family, and that no-one else in the household spoke anything other than the local dialect. I learned some German—but after a few weeks I took off on my bicycle again, this time by myself, and crossed Switzerland as far as Lausanne, where I joined up with some friends and somehow gained admission to the meeting of the League of Nations on the occasion when Germany was admitted. Briand, from France, threw the machine guns behind him in his speech; Stresemann was less dramatic and incomprehensible to me. I managed to sell my bicycle in Geneva and came home by train.

Of my undergraduate Cambridge friends, I have only kept up a friendship throughout life with Frank Oldham. I remember my first year, particularly, as crowded with the new experiences that I have described. But in my second year I began to ask myself—how am I going to establish myself as someone who has a contribution to make in the community? It seemed to me that this could only be in science; I would have liked to make a mark in the Union Society, but at that time I had not the necessary self-assurance to raise my voice there. I had—so I felt—no other talent except mathematics. Also I became dimly aware that a revolution in physics was on the way; I was in a hurry to take part in it. In my second year, work for the examination already assumed a new urgency; I simply had to prove myself. I am sure I continued to enjoy Cambridge life and my friends, but I cannot remember much of it. What I remember was the desire to succeed, in the only way I could. This I have kept, all through my career. However, an extract from a letter written to my fiancée in July 1928 when she was away in Germany gives perhaps a different picture.

> Being here, with you away for so long, makes me realise how much things have changed in the last few months. Now I'm going to talk about myself!! I do feel that till now I've lived a trivial, detached, rather lazy life. Need I make excuses? I had a dreadful time at school. I've never told you, or anyone else. I always pretended to my parents that I was happy, in case they should worry about me. So till I was 18 I had no friend really—I was perfectly alone. So merely in self-defence I had to make myself so that I didn't mind things, so that I could laugh at things. But since I left school life has been great fun; a book, a friend, a tour across Europe. And work—that was interesting so that I worked hard. The thrill of my first paper, of having added a little bit to knowledge, was enormous. I curse myself for not having done lots of things—public speaking, learning a musical instrument, some psychology. But I never cared enough to make the effort. And I never cared what was going to happen to me, so long as I was amused. Sometimes I had a dreadful feeling of pointlessness (most people do).
>
> Well, now it's all quite different. I would like to do something really worthwhile, and it is perplexing, as I don't know what it is. The pathos and waste of the world are so terrible . . .!

The next three letters were written to my parents in 1926.

21.2.26

In your day, did the 'Universe' lose interest to some extent after one's first year? One tends to get specialised. One can with an effort recall the facts that right and wrong are subjective, ..., that morals are a question of aesthetics. One can even remember why. But the old fluency has gone. It's the same with everyone, almost proverbially. At a discussion group that I turn up to occasionally, most of the talking is done by the freshers. Last time, a thoughtful fresher remarking that the existence of a God precluded the universal applicability of the law of causality; Harper and I spent the next half hour putting that into symbolical logic. We succeeded,

$$\therefore \quad \{(\exists p) \supset :(\exists q)q \supset p\}/\{(\exists v)v. = .\hat{x}(x = x)\}$$

God is the all inclusive class; i.e. the class of all x for which $x = x$.

The proposition cannot be proved.

No, we are not always so childish.

But when freshers talk about God, then second year men think 'I know all about that.' When I was a fresher, I discussed it and came to a perfectly satisfactory conclusion.

But one forgets what the conclusion was.

4.3.26

I am trying to write the minutes of an imaginary study circle. At one of ours, we discussed whether the marriage laws wanted changing. At the end someone solemnly remarked that we ought to meet again in 20 years time. I am writing the minutes of the future meeting. Great fun!

Religion perhaps then played a greater part in undergraduate life than it does today. Like other undergraduates, I often went to meetings in my first year to hear different views.

After a talk with some CICCU (Cambridge Intercollegiate Christian Union) people, I went to one of their meetings and was repelled when they prayed from the platform for a 'man in John's' (myself) 'whose heart was moved . . .' I don't think I had much interest. I wrote to my mother.

1.5.1926

I've been busy this week with this Cambridge Inter-collegiate Christian fellowship (C.I.C.C.U.), Student Christian Movement and League of Nations Union, all getting busy arranging programs for next year, finding college representatives and leaders for study circles, etc. At two meetings of the two former we had rather heated arguments between the people who wanted to emphasize the 'C'

(Christian) aspects and prayers at study circles, prayers everywhere and other things—and the people who didn't, which seemed to be mostly me! If the S.C.M. gets too Christian, it will go out, because the whole-hogging enthusiast goes to the C.I.C.C.U.; the people who make the S.C.M. go are the scientific enquiring people, the people with ethical 'C' ideas. But the people who are C but not C enough for the C.I.C.C.U. will never keep it going.

Another letter is undated, perhaps a year or two later:

Dearest Mum,
Sir James Jean's book is exciting. [*Astronomy and Cosmology*, 1928]

Supposing one considers that the stars were created in order that there might be *life* on just one or two of their planets—it's like the million herring's eggs from which grows up one herring. It would seem that the creator, wanting to create man, forswore any such method as that of Genesis, and created just enough matter for a home for man to be bound to evolve out of it at last. And then when the home has been made, man is made by rather similar methods,—at any rate, a method in which nearly everything is waste. If one thinks that man is the centre of the universe, then Astronomy just makes one unimpressed with mere size. The vastness of those incredible great furnaces, 100,000 light years away, 40,000,000 degrees hot etc.—must be rather less commanding of respect than the circulation, 1,410,176 net sale of the Daily Mail.

Of course one can say that man is an impudent little speck dwarfed by all these things. But I don't see what size has to do with it—except as showing Nature's lavishness.

But Quantum theory makes one believe very much that what is external, and 180,000 light years broad, and 100,000,000 years old, is only a relationship between man's possible sensations —, a kind of framework; and that does make life much the most *real* thing.

I think it probable that quantum theory will eventually be able to predict the number of electrons that there must be in the Universe.

I like to think that the deity is to man as man is to matter, so that just as there can be matter but no life, there can be man but no god—God being something more complex growing out of man's inspirations—.

But there must be another God, God the architect (who by the way has no 'sense of time', I think).

Forgive five pages of perfect nonsense.

I was an undergraduate for one more year, till June 1927, preparing for research and with all examinations behind me, as I describe in the next chapter. I include extracts from two further letters to my parents, one before and one after the long vacation of that year.

May 1927

Everybody's last term. We promenade the backs in the evenings about 10 strong. Our set. But talking of sets, I think that scarcely any of our set did I know during my first year. Then we were all just clever young men. Clever and less clever. Horizontal divisions. Now divisions are vertical; people 'qui sont sympathiques'.

27.11.27

What shall one write about, on a cold Saturday evening so near the end of term? Work? I've a little note book with thirty or forty nicely written pages—the product of six weeks pretty hard work. I'm trying to write it up a bit, so that Dirac can see it—and all that stuff takes a lot of arranging, before anyone else could make it out. I took it down to Fowler and showed him the results I had. This was this morning in the Cavendish Library. He thought there was something in it—really didn't know much about it; that is the penalty of working out one's own bright ideas, instead of doing the dull things that Fowler suggests! But Dirac was there, and Fowler called him and Dirac said timidly that it was all nonsense, and referred me to one of his papers—which is about something quite different. Dirac said that the general theory allowed us to assume ..... I asked him how he knew, and because I thought that the great man was being stupid, I may have summoned up courage to hector the great. Then I suddenly realised that the great man was timid and that I was being a bully'! Funny moment. Fowler suggested that I should write it all out nicely and that Dirac should read it and Dirac said he would—I hope he won't hate me too much!

However, if I can work out a little integral, get a second term in an expansion, then I shall get out my log tables, and play about with *real numbers*, and go down to P.M.S. Blackett in his lab, and say 'Do the alpha particles go just so? Is the deviation from the Classical Law just this?' And the final arbitrator will speak from out of his glass tubing and say Yes—or No. And if he says Yes, my reputation is made, because all the people in Germany have been trying to get this thing for quite six months. And I shall take a holiday. But Dirac says my method is all rot, and perhaps I can't work out the integral, and perhaps the oracle will speak out of his lab.

and say no. In which case I can merely hope that my next effort will be more successful.

Well it sounds awfully exciting doesn't it? As a matter of fact it is rather!

It would spoil the effect to say now what it's all about.

Although I cannot remember now what the problem was, this particular piece of research was not, I believe, successful.

CHAPTER FOUR

First Research

Returning to Cambridge in October 1926, still as undergraduate, John Brunyate and I were all set to start research. I think that 1926 was the most fortunate time to start research in theoretical physics because there were so many easy problems to solve. The situation was as follows. Ernest Rutherford in Manchester in 1905 had shown that the atom, hitherto thought of as a hard little billiard ball, was in fact a miniature solar system, with a heavy charged nucleus and a number of electrons revolving round it. The lightest of the atoms, hydrogen, had only one electron. But the laws that governed the motion of these electrons were not known, nor was it known why the atom kept its shape and size when continuously bumping against other atoms. Just before the first world war, a young Danish physicist, Niels Bohr, working in Manchester at the time, produced some quite revolutionary theories limiting the orbits in which these electrons could move. This was the Quantum Theory, and it could explain a lot of facts. A year or two after the 1914–18 war, it was fairly widely accepted and these orbits were in all the textbooks. I knew about all this in my school-days, and still have books illustrating these orbits which I acquired as college prizes (e.g. E. N. da C. Andrade, *The Structure of the Atom*. Some of his pictures of orbits are illustrated in Figure 4 (he must have believed in them devoutly). But—more and more—it became clear that though Bohr's theory could explain many things, it couldn't explain everything. Something quite new was needed, a radically different system of mechanics for electrons in atoms to replace the Newtonian mechanics that worked so well for the planets, and which had only been tinkered with by Bohr. Suddenly it came; a young German physicist working in Göttingen, Werner Heisenberg, was the leading man. He used a form of mathematics with which most of us in Cambridge were not familiar called matrix mechanics. And very shortly afterwards Erwin Schrödinger in Vienna produced an alternative form of the theory which used differential equations, a much more familiar technique. This was called wave mechanics. It was accepted in the inner circles of theoretical physics, Göttingen, Copenhagen and

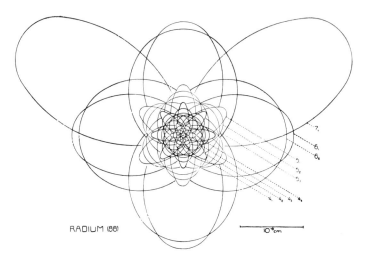

RADIUM (88)

10^{-8} cm

Figure 4. Structure of the Bohr atom for radium.

Zurich within six months. It meant that once one understood the new theories and combined mathematics and intuition, one could discover why nature worked as it did for the great majority of the problems of physics and of chemistry.

In Cambridge the only man who could understand the new theories, and who indeed did much to develop them, was that remarkable genius Paul Dirac† (see letter in Chapter 3), and it was hard to talk to him. A colleague sitting next to him at dinner is said to have asked him what he was working on and got the reply 'Do you know what adiabatic invariants are?' He replied 'No' and was silenced on hearing 'What is the use of my talking to you if you don't know the very elements of the subject?' In charge of theoretical physics was Ralph Fowler, Rutherford's son-in-law. Brunyate and I went to see him, hoping to get a problem to work on. But he was off to America for a year's study leave, and told us to come back when he returned. This was too much for Brunyate, who gave up physics and took to law, eventually becoming solicitor to the university. For me it meant, as far as work was concerned, a rather solitary year, reading the original papers, which apart from Dirac's were mostly in German. I think I found the German harder than the physics, especially in Schrödinger's papers. Brunyate had not welcomed these, being delighted by the unfamiliar mathematics in

†Dirac was born in 1902, FRS 1930, Nobel prize 1933, Lucasian Professor of Physics in Cambridge 1932–69. Died 1984.

Heisenberg's work, and feeling that differential equations were old hat. But my ambition was to prove myself and to explain something, and I welcomed the more familiar mathematics of Schrödinger.

It was a year before I succeeded. Rutherford then bestrode British physics like the giant he was, and at the back of his work was the 'scattering law', the law giving the number of fast-moving particles that would be turned back by the charged atomic nucleus when they were directed onto a thin sheet of some materials. This had been deduced using Newtonian mechanics in Rutherford's Manchester days. But, following Schrödinger, we knew that beams of particles had to be treated as if they were waves; the law had to be deduced from Schrödinger's equation. I was able to give the first proof. I remember vividly how, one afternoon, walking back to my lodgings, I realised how the proof must go. I don't think this made a great impression in the Cavendish; one could have guessed that it must be so, but Fowler, now

Figure 5. The Old Cavendish, seen from Free School Lane, Cambridge.

Figure 6. The practical physics class in the Cavendish laboratory, showing Dr G.F.C. Searle.

back from America, thought well of it and sent it to be published in the prestigious *Proceedings of the Royal Society*. I felt that I was on the road to being a physicist, especially when people wrote and asked for reprints.

At that time theoretical physics did not have much place in the Cavendish. Fowler had a room, but for students there was nowhere to sit, except the rather small and squalid library, which also served as a tea room. We were in the Faculty of Mathematics, and the tradition was that we should sit in our college rooms and think. The fact that frequent discussions between theorists and experimentalists were essential was something I only learned later.

Rutherford was said not to be friendly to theory, except perhaps when it came from Bohr, though later he saw how quantum mechanics opened the way to our understanding of the nucleus and the atom, and welcomed it. If I had been an experimental worker, I should have seen him, Chadwick and the other famous men quite often, walking round the laboratory to see how the experiments were going. But, as I have said, we theorists were left very much to ourselves to sink or swim—or at any rate I was. But I have no complaints. Once I understood the new theories, there were *easy* problems to be tackled. I think a decade later, when the easy ones had been worked out, I might have foundered with so little help.

It was during my first year after graduating, in the spring of 1928, that I became engaged to Ruth Horder; we married two years later. Ruth was the daughter of Gerald Horder, a quantity surveyor with a beautiful old house in Ealing. Her uncle, Percy Morley Horder, was a distinguished architect. She had won a scholarship in classics from St. Pauls Girls' School to Newnham College, Cambridge—and I first met her at a tea party in that college, invited by a friend of my sister, who came to Newnham in the same year. A woman's college concerned itself with such matters in those days, and she and I were asked to see her tutor. It was explained to us, in the nicest possible way, that while we could of course go for walks together, this was not so after dark. I think we were amused. She obtained a first in part I of the classics tripos and, after an extra year a first in part II, taking classical archaeology as her special subject.

CHAPTER FIVE

Copenhagen 1928

I was extremely fortunate to be able to go abroad one year after graduating, financed by a grant from the Department of Scientific and Industrial Research (the equivalent of what is now the Science and Engineering Research Council). R. H. Fowler arranged it, and I am deeply grateful to him for this as for so much else. I planned to spend the autumn term in Copenhagen with Niels Bohr, the Lent Term back in Cambridge (my fiancée being still at Newnham) and the summer term in Göttingen in Germany.

In an earlier chapter I have referred to Bohr's quantum theory which he presented to the world just before the first world war. His first assumption was that the energy of any atom or molecule was 'quantized'. This meant that it behaved quite differently from—for instance—the solar system, as described by Isaac Newton in the seventeenth century. If one imagined some body from outer space moving through our solar system, it would leave it slightly changed; days might be longer or shorter than before, for example. But atoms or molecules—according to Bohr—were not like that; you could hit them quite hard, with for instance a beam of electrons, and nothing was changed. But if you hit them hard enough, they went into what was called an 'excited stationary state', where they stayed for a while, and then gave out light, returning to their original state. All this part of Bohr's quantum theory has survived to this day, and there is ample experimental evidence for its validity. But in addition Bohr wanted to calculate the energies of these stationary states, which could be deduced experimentally from the wavelengths (colour) of the light emitted. He therefore proposed the orbits which are illustrated in Figure 4. The 'new' quantum mechanics of Heisenberg and Schrödinger in 1925–6 completely replaced these ideas. Everyone recognized this at once, including Bohr. I am sure he made no attempt to defend his orbits—as so many of us do with our brain-children, when the advance of science shows that they must be replaced. But Bohr, through his genius, his personality and through the respect and affection everyone had for him, had kept Copenhagen as the centre to which everyone came, to put their ideas before him and hear what he had to say about them.

Just before I came to Copenhagen, Paul Dirac in Cambridge had startled the small world of the atomic physicists by what I believe to have been the most brilliant piece of reasoning that I have seen in my lifetime. It had been known for some years, through examination of the radiation (light) emitted by various gases, that the electron, the fundamental particle of electricity and a constituent of all matter, was not just a point charge, but seemed to have an axis of symmetry. One had to ask, not only where it was in space and how fast it was moving, but in what direction this axis was pointing. We spoke of it as the 'spinning electron', as though it were some kind of top. It seemed strange that nature should need so complicated a form for its most fundamental of particles. But Dirac showed that, if nature obeyed the quantum mechanics which he—Dirac—had done so much to clarify, and also Einstein's principle of relativity, the electron *must* have a spin; and he produced his famous wave equation which made it possible to calculate its properties in detail.

When I came to Copenhagen I was already working on the consequences of Dirac's theory. My first research had been on Rutherford's law for the scattering of alpha particles or electrons by atomic nuclei and I wanted to see what changes the new theory would bring about. It was of course the obvious thing for me to do, given the experience I had already and my intense desire to establish myself properly in my own small world. Most of my discussions with Bohr were about this problem, and particularly whether this spin could be observed in a free electron, or only when the electron was in an atom. Bohr seemed to think it could not in the former case. I do not think that these discussions had any profound effect on the advancement of science—but I learned what physics was all about, how it was a social activity and how a teacher should be with students and how beautiful physics could be. For us theorists the whole atmosphere was as different as it could be from that in Cambridge. We came to the institute at ten and left about six. We were in and out of each others' rooms all day and so was Bohr. No-one would ever keep an idea to himself. In fact I might say that he gave to us theoreticians a life such as Rutherford created for his 'boys', the experimentalists at the Cavendish.

As regards discussions with Bohr, I cannot do better than reproduce some letters to my mother.

10.9.28

It was quite thrilling at Bohr's on Saturday evening. Lots of people there; Mrs. Bohr, and the two eldest of the children, handing round delicious eats and drinks. Pauli sitting on a chair and rocking his fat body about and telling humorous Jewish stories. Gamow talking about Russia. And Bohr going round and talking shop to one person privately after another; fearfully eager looking, asking about one's own work, glowing when he talks of the big problem yet to be solved, that is to be discussed when Heisenberg comes here. That is the problem, perhaps, of the interconnection between Relativity and the Quantum Theory, which contradict one another rather.

23.9.28

I like to hear Bohr when he has been shown a new theory. This was part of the radio-active theory, which is very simple and obvious, once one has grasped the principle of wave mechanics. Bohr looks at the theory for a long time in silence, his face full of wonder and admiration, as before a great work of art. 'Schön, wunderbar schön' [beautiful, wonderfully beautiful] he says slowly. Then he turns to the Lord God, the ineffable mystery, and, with the greatest reverence; 'es ist ein Wunderbares Ding, die Wellenmechanik' [it is a wonderful thing, wave mechanics]. But you should hear the way he says it.

I have been able to have several talks with him, and told him something that he did not believe; so now I must prove it!

6.10.28

Yesterday at four o'clock Bohr said, 'come across to my house and discuss the little bit of work that you have just done, and a bit of work rather similar of someone else's'; Bohr lives opposite the Institute. And at six it was supper time, and Bohr said stay for supper, and I had supper with him and his wife, and we talked about sculpture. And then after supper we went on discussing, and it became more and more Bohr and less and less me. And by about 9 we had got about as far with the problem as seemed possible without further calculations, and so Bohr began to talk about the Philosophy of the Quantum Theory and how it was all bound up with the impossibility of man's knowing himself, and his not being able to know the external world completely because he himself was a part of the external world. And then back to the Quantum Theory and the outstanding problems again. And about eleven we said good night.

It is incredibly nice of him, isn't it, to give individual students this attention.

He *has* got a brain. When he has a new idea—he had this

morning—he comes into the Institut and tells it to the first person he can find; today that was me.

Extraordinary what a difference it makes to life in general if one's work is going well. When I got messed up in that beastly arithmetic, and the thing gave an answer that didn't agree with Rutherford's experiments, I felt that I was stupid, would never do any good at this game, consequently rather overworked, and got dreadfully fed up and wondered if I'd better not come home and get a job in business like [my cousin] Johnny Simmonds. But now all goes well.

I like the life here where half one's work is discussing. That is the great point of Copenhagen I believe—though its probably the same in Germany too. Only Bohr knows *everything* that's being done, and has a marvellous knack of finding the sense behind mathematics.

Bohr is the kind of man who can tell one that one is talking nonsense, without hurting—I don't think one can say more than that, do you?

And he has his students alone in the evening to talk, and then walks home with them, telling how he discovered his theory of spectra.

And then its 1 a.m. perhaps.

But it is funny that the spin of the electron can never be observed, isn't it? Perhaps the spin is only an illusion.

11.11.28

Now I am beginning to wonder when to come home. I think I must just say, when I have finished this spinning electron problem that I am doing with Bohr and Klein. I hardly can leave in the middle can I?

But perhaps we shall get a result soon.

I do most of the mathematics, but Bohr provides the ideas! It is the most thrilling problem I have ever been working on; something rather fundamental, about whether the electron 'really' has a spin or not.

It is so exciting—I never want to stop trying it in the evenings till I simply can't any more. But it is rather obstinate.

And it'll be rather extra lovely coming home if that is solved—!

December, 1928

Niels Bohr, one realises more and more, is almost a national hero here in Denmark. Well, Denmark—that is Copenhagen; the rest is peasants and country towns. But still, he is infinitely better known than Rutherford, for instance, in Cambridge. There is a story that a young man arrived in Copenhagen and took a taxi to Bohr's institute; and the taxi man wouldn't take any money because it was to Bohr's that he had driven—and the veriest business man knows all about him.

Whether it is that Denmark is proud to have such a great physicist, or whether it is that he had done such a lot of other things, raised money to buy radium for the hospital, or to build a new museum,—I don't know.

And he is *such* a charming man.

The other day he rushed away from the Institute for an hour, with Gamow and Hartree and me, all in a taxi. And he took us round this museum—it is Danish primitive man, and the best in the world they say; and what a guide ...!

And Gamow—Gamow is a Leningrad student, and it is not so easy for Russians to get a visa for Denmark, but when he said, to visit Professor Bohr—all easy. And Gamow too, talking about his lodging house—he comes in late and says he has been talking to Bohr; and the old gentleman at the end of the table sits up with a jerk—'Herr Professor Bohr; these young men have no reverence'.

And today he told that he was going for a walk in the country with Bohr 'Sie gehen auf dem Land mit dem Herrn Professor Bohr—!—By Jove you *must* be somebody'.

And its so nice, any stray person one meets in the tram or anywhere, if you tell them you are working under Professor Bohr, they know all about you at once.

Tomorrow Gamow and I are going for a day into the country with him.

And this is the man who will walk home with one from the Institute, sometimes later than midnight, right up to the door of Livjaegergade and past it and back again, telling how he first thought of the quantum theory.

November, 1928

Gamow is a pleasant lively young man at the Institute who has a most ingenious theory about the radioactive nucleus. Though he is a Russian, one wouldn't think it; he is the Oliver Walker type, goes to the cinema rather often, and would love a motor cycle if he had one. And he reads Conan Doyle and doesn't go to concerts, but is a brilliant physicist and hard working, and gets his results without using mathematics. And he very seldom stops talking and is about my height.

Does this introduce him?

He tells stories about the Revolution. Of course he was 16 or

17 at the bad times; I suppose he was short of food like anyone else, but he doesn't mention that. It all sounds such fun, as he tells it; not that it could have been.

He lived in Odessa; White Army and Red Army fighting in the streets. Staying in his house was a cousin, recently married to an officer in the White Army; the White Army was going to evacuate that night, and she must go and join them before they went. Streets dangerous. He, 16 years old, said he would go with her to the White Army; but mustn't tell papa, he wouldn't allow it! Crept out behind through the kitchen door, got through the streets safely, then he went back, expecting any moment a burst of machine gun fire up the street.

Since then they've never heard of their cousin. Perhaps she lives, perhaps she does not.

And so—Motti—what a pity you were not there.

Gamow was my closest friend in Copenhagen. We went to the cinema, talked shop and everything else. He borrowed 25 öre from me most days to buy cigarettes. He had already had great success, having shown how the new quantum mechanics could explain radioactivity—how an atom could be dormant for millions of years and suddenly break up. I was—I must say—rather jealous. 'Ah Motti'—he used to say—you must construct an alpha particle—meaning, to make a theory of how it hung together. Of course I couldn't do that; no-one could attempt it till the neutron was discovered in 1932. My first important paper came when back in Cambridge in 1929, and was on quite a different subject.

Gamow left Russia in 1933 never to return; he had a distinguished career in the USA, working in atomic energy and writing popular books about science. A short autobiography was published after his death in 1968[1].

My memory of Copenhagen, however, is most vividly of Bohr, taking my ideas, examining them from every point of view, walking round the table first one way, then the other, week after week. Perhaps I was impatient; I wanted to get them into print before anyone else had a similar idea. But it was a marvellous education.

CHAPTER SIX

Cambridge and Göttingen 1929

In Cambridge next term (January 1929), I remember Patrick Blackett (later Lord Blackett and President of the Royal Society) saying—'you are a changed character, Mott, since Copenhagen. You are in the lab. all day!' And it was then that the work occurred that really established me as a physicist. It began in Copenhagen. A paper by J. R. Oppenheimer (later famous as a leader in the atomic bomb project)showed that, in collision problems, when electrons hit atoms, quite large effects dependent on the spin direction could occur. I thought of applying this to collision between free electrons—another problem that Bohr and I went round and round. But Fowler suggested that I apply my ideas to the alpha particle, which hasn't any spin at all. So I did this. Now the alpha particle was identical with the nucleus of the helium atom. I was able to predict that, if alpha particles traversed some helium gas, the number deflected through $45°$ would be twice as many as predicted by the famous Rutherford scattering formula. This was because the two particles were identical and after the collision one couldn't tell which was which. It was in fact meaningless to ask. But the experiment had to be done for very slow particles because it was known that for helium, where the charge is comparatively small, fast particles would come so close that the (unknown) structure of the particles would play a major role. That was why the effect had not been observed; the interest was focussed on this unknown structure and thus on fast particles. James Chadwick, Rutherford's second in command, undertook to see if the effect existed. He had to slow down the particles; they came from a radioactive source and were passed through thin foils of lead before entering the chamber where the helium was. He showed me—perhaps a year later—the results, which I reproduce in Figure 7. The intensity of the scattering is plotted agains $1/V^2$, where V is the velocity of the particles. The dotted line represents the number of particles deflected as predicted by the Rutherford formula. The effect was there all right. Chadwick took me along to see Rutherford, who said 'If you think of anything else like this, come and tell me.' This certainly made my day. In fact

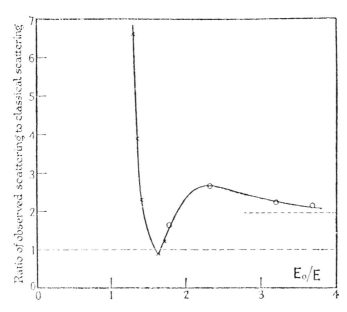

Figure 7. Scattering of alpha particles by helium at angles between 40° and 50°. (J. Chadwick)

I think it was on this day that I gained complete confidence in my ability to make a career in theoretical physics, so that I could cease to worry about it.

After Easter I went to Göttingen. Göttingen was the place where, above all, the new quantum mechanics had originated. I think I chose it partly for that and partly because I was fascinated by Germany in the days of the Weimar Republic. But I would have done better to go back to Copenhagen. I was supposed to be under the senior professor there, Max Born, but he was, I believe, in poor health and I cannot remember that I ever saw him.

Although Heisenberg had left, there were still good people around, but my German, which had been good enough for the polyglot discussions in Copenhagen, did not seem adequate for me to push myself into this society where one sank or swam. I made one German friend, Martin Stobbe, but on the whole I was rather solitary, going off on a bicycle into the Thüringer Wald, staying at small inns where a bed could be got for 60 pfennig. I remember one with pictures on the wall on either side, one of Jesus Christ and one of Kaiser Wilhelm the second. The atmosphere is perhaps suggested by these extracts from a letter (1.6.29):

Figure 8. The author when a research student at Cambridge.

This, I believe, is where Victoria used to go with Albert. Anyway, I went through Gotha, where there was a Residenz, and that surely was where Albert used to live. And didn't Albert have Heimweh for Gotha and the Thüringerwald?

The Thüringerwald is very nice, pine trees and a fine mountainous horizon. Gotha, that Albert was prince of, is a little town like Freiburg, and not a bit grand. But perhaps Saxe-Coburg is.

And in a shop where I said I was English—they suggested—
'Oh, I suppose you know Albert then'.

Later Ruth came to join me, and we stayed with a retired Admiral and
his wife, and later went to a conference in Zurich. The following letter
describes the Admiral:

June 1929

It is quite true that the Germans eat too much—some Germans,
at any rate! Though they seem to flourish on it even if they do get
a bit fat! In this sweltering weather the Admiral and his wife put
away, at 1.30, soup, more meat than any of our family ever eat at
any meal in the Lake District, and a creamy pudding and bread and
cheese. And poor Ruth and I would just like a little tomato salad,
perhaps! They are quite hurt—können die Engländer nicht essen?
Queer! I wonder how they manage it.

Otherwise they are most charming, and I have learned more
German at meals with them than in all the time I've been here. I
shall feel quite competent to give a lecture at Zurich. The Admiral
of course was rich before the war and knew the Kaiser. I have never
dared to ask him if he was in any battle. Naturally, I've never heard
a word from him against any of Germany's late enemies—but lots
and lots of words against the German republic! There is a German
war book *Nothing new on the Western front—Im Westen Nichts
neues*—you may have heard of it. Ruth and I have been reading it,
and we think it amazingly good. But there's precious little about
honour and glory and das Vaterland in it—precious little! And
rashly we told the Admiral we were reading it, and he didn't like
it a bit that we were reading it. Quite a wrong impression of the
war, he said, and he recommended me another all about dying for
the fatherland.

But they are charming. To see the Admiral kissing a lady's
hand is quite beautiful. And he has collected butterflies all over the
world, all the places he went to with his ship. And we went on an
Ausflug [expedition] and the Admiral took his net, and chased
them like a little boy.

At that time I had to decide what to do next year. I had plenty of
offers—back to Cambridge, Bristol, Zurich (with Pauli). There were certainly more jobs in physics theory than people; in the UK there were
practically only myself and Alan Wilson, with Dirac and Douglas Hartree a few years older. Eventually I was offered a lectureship with tenure
at Manchester, under W. L. Bragg, at £400 a year. Cambridge and

Manchester were the only worthwhile schools of physics in the UK (with Bristol just beginning) and I liked and admired Bragg, so I accepted.

Ruth went back to Newnham for a fourth year, to read classical archaeology which suited her much better than philosophy. We planned to get married in the summer of 1930, with parental blessing on both sides—though in the event we married rather earlier.

CHAPTER SEVEN

Manchester 1929–30

My year at Manchester under Professor Bragg was most rewarding, and I regret in many ways that I did not stay longer.

At the age of 25 W. L. Bragg, later Sir Lawrence, was awarded the Nobel prize for physics, jointly with his father W. H. Bragg, for their 'services in the analysis of crystal structure by means of X-rays'. In fact, they were the originators of a new science, the scientific investigation of the structure of crystals, and that meant of most materials. Since the end of the war Bragg had held the chair at Manchester and was the world leader in this subject. It was a change from Rutherford's Cavendish and indeed Bohr's Copenhagen, where the atom, the nucleus and the electron were the thing. Here one asked, and found out, how atoms were put together to form real materials. It was an introduction to 'solid state physics', in which I made my career from 1933 onwards.

Why did I go there? Partly the offer of an established position with tenure, and £400 p.a. was a very tolerable salary. Then there was the prestige of the position; I could think of it as the one Bohr had held, theorist to the Manchester physics department. Partly liking for Bragg—and also the urgings of Rutherford that no-one should stay in Cambridge too long. He told me, (according to a letter to my mother):

> 13.1.30
> I always believe in young men leaving Cambridge; I find that after about five years in Cambridge they get a sort of paralysis of the will, don't you know. I've had to send away a lot of young men from my lab. Now I went to Canada when I was 27, and *I've* stood on my own feet ever since.

Fowler wrote,

> Dear Mott,
> I am glad to hear about Manchester. That is quite right! Don't go to sleep there, but make up your mind to get on with the job of collision problems however much else you have to do.

Let me see what you get out when its ready.

Good luck for your researches into physical and human equations.

[The latter refers to my intended marriage.]

My main job was to give a course of lectures, attended by the professor and most of his staff, to explain what wave mechanics was all about—quite a job for a young man of 25. However I survived and this led me to write my first book, *An Outline of Wave Mechanics*[2]. I also did some student teaching, including, for the only time in my life, work in the practical class. Also the ideas of crystal physics were opened up to me. My feelings about physics are described in a letter to my mother, at about the end of 1929.

Ancoats

I am very busy at the moment; at Manchester one does not think of less bright ideas than at Cambridge or Copenhagen; but there is less time to work them out. I have a very nice one at the moment, and am looking forward to having a good slog at it at the weekend. It is the only opportunity for uninterrupted work.

It is nice having one's room in the physics department. People are always coming in and asking about something, either some point connected with my lectures or something connected with their own work. One of the older research students said the sweetest thing to me after my lecture, that he had never realised that there was anything aesthetic in Mathematics till one of my lectures. I was frightfully bucked.

Physical theories have got a frightful fascination for me. I used to think it was just the pleasure of 'getting something out' that made me like research. I expect at first it was. But now that I have to review the whole thing in my mind, and read up a lot of aspects of it that I had never considered before, I find a great deal more in it than that. Rather the same pleasure that there is in a picture with a beautiful shape—the fascination of that Ethelbert White for instance [a picture the family had bought]; only that strikes one at once and holds one, whereas the shape of this quantum theory is all hidden and has to be delved for.

Doing research must be awfully like digging up the foundations of an old temple. Only what an architect it was that architected the Quantum Theory.

Introspective digging.

This seems to be an introspective letter.

The regrettable thing about me is that I find it so infinitely more interesting than the position in Kenya, about which we have had a lecture here tonight.

During Ruth's fourth year at Newnham, her interest in Anglo-Catholicism provoked the following letter from me at the end of 1929:

> Dearest, when you say the Catholic church in your letter, you mean always the Anglo-Catholic, don't you? When you say Catholic I always shudder slightly, because Catholic usually means Roman Catholic. That makes me feel cold all over. If you *wanted* to be a Roman Catholic, I should want to die. The trouble is that I can't think dispassionately about the institution because it fills me with such absolute physical disgust. That church with the beautiful Gilbert Scott tower near the university, I have been into it once. The vague smell of incense, the tawdry madonnas with their candles, St. Peter with his shiny toe, the black muffled priests, the well-dressed women with blank faces who dip their fingers in holy water, the confessors' boxes...

I need hardly say that I do not feel like this now. We have some very dear Catholic friends, we know the devoted service Catholic nuns can give to the handicapped and as, I relate later on, I have been much influenced by the theology of Hans Küng, whom I know personally. But—none the less—something of my antagonism remains, and I would be sad if any near relation joined that church.

My living conditions were almost as exciting to me as my physics. I lived in the Manchester University Settlement, in Every Street, Ancoats, within walking distance of the University. It was housed in two rather beautiful eighteenth century buildings, in a picturesque but decaying street of working class housing. The warden, Miss Cashmore, was a remarkable woman, who made the Settlement a centre for anyone interested in the social conditions in Britain at the beginning of the great depression; staff of the *Manchester Guardian*, then edited by C. P. Scott, used to drop in to lunch. But fascinating above all was the interest taken in the Settlement by Mary Stocks (later Baroness Stocks) and her husband, Professor of Classics and later Vice-Chancellor of Liverpool. Mary Stocks was a friend of my mother with similar interests on social matters. Almost certainly it was she who told me of the Settlement. She wrote a miracle play, *Every Man of Every Street*, for performance by Ancoats people and university teachers; I was the old king, presenting gifts to the child Jesus. Meeting people of such different backgrounds was something new to me.

As well as being near the university, Ancoats was near London Road railway station and within an hour I could be in the Peak District, walking over the green hills criss-crossed with stone walls, like those that I remembered from my childhood.

In the spring I got invitations to return to Cambridge, not from my old college (St John's), but from King's (promoted by Patrick Blackett) and from Gonville and Caius. In those days about half a lecturer's salary would come from the college, for which he would be expected to give individual teaching in mathematics for about eight hours every week, and the rest came from the university for formal lecturing (six hours) and examining. I loved Manchester—but there were many reasons for going. The total salary was twice as great as at Manchester and I was about to marry. I did not expect to stay very long at Manchester; to get a house or flat and then move would be disturbing. The prestige of being a Fellow of a college was considerable; and the Cavendish was at the very peak of its reputation, perhaps the most famous physics centre in the world. I decided to accept. I would have preferred King's, because of my admiration for Patrick Blackett, but the invitation from Caius came first, and I accepted. Bragg and his colleagues were sorry but felt that I could not do otherwise.

A few years ago I was in Manchester and asked to be driven round Ancoats. Ancoats Hall was still there, but every street around it had disappeared, to be replaced by parking lots and waste land. I do not know what has become of the Settlement.

CHAPTER EIGHT

Cambridge 1930–33

Ruth and I, now married, bought and furnished a house in Sedley Taylor Road in time for the autumn. We bought a car, a Wolsey with two cylinders, and a dog; the dog, alas, was run over by a train and we never had another.

The three years during which I was in Cambridge included 1932, the *Annus Mirabilis* of the Cavendish, when the neutron was discovered, when the accelerators were first able to 'split the atom' and where the positron—or positive electron—predicted by Dirac was first observed. It was hardly the *Annus Mirabilis* for anyone else, with the industrial depression deepening and the Nazis going from strength to strength in Germany. We were intensely aware of that—but aware at the same time that we in the Cavendish were part of a vastly important moment of history. A guest at Trinity College is said to have told Rutherford after the discovery of the neutron, 'Lucky man, Rutherford, always on the crest of the wave'. He got the reply, 'Crest of the wave, indeed. I have made the wave, didn't I?' And we all felt that we helped to keep the wave moving.

The Cavendish seemed now to give a much greater role to theorists. Fowler was now Professor of Theoretical Physics, still with a room there next to Rutherford's. German physicists of my generation, such as Hans Bethe and Rudolf Peierls, were in and out. Paul Dirac seemed much more approachable. But—apart from Fowler—we theoreticians had no official position there, nor anywhere except the library in which to work. I was a member of the faculty of mathematics; one course of lectures was for Part I of the mathematical tripos, and was rather dull; the other, in the Cavendish, was on quantum theory and this I enjoyed, though in the first term my audience melted away! In Cambridge, where attendance at lectures is not compulsory, a lecturer has to learn his trade. I did not very much enjoy the eight weekly hours of 'supervision' at Caius. At that time the solution of quite difficult and rather artificial questions was a major part of the mathematical tripos, and I found them increasingly difficult to solve for my students. With my increasing involvement in physics it seemed a waste of time.

Caius seemed to me a rigid and traditional society. As junior Fellow I had the duty of passing round the port, and recording any bets made by my colleagues. College records tell how much port was drunk when the first news arrived of the victory at Trafalgar in 1805, before the death of Nelson was known. We were more moderate in 1930. Of course we were strictly a single sex college and no woman could be invited to lunch or dinner, or to the Fellows' Christmas party where there was a lucky dip and some kind of balloon football. With a newly married wife and with all that was happening in the Cavendish, I was rather lukewarm about all this.

Fowler once again went away on sabbatical leave, and asked me to look after his research students. They included Harold Taylor, a lifelong friend, who later took high office in the university administration. I had to find problems for them, which was not difficult, since so much was going on among the experimentalists. Since I had never taken my Ph.D. and so was paying £4 a term to Fowler as my supervisor, at the same time as I was receiving fees from Fowler's students, it did not seem worthwhile continuing to register myself as a candidate; I thought a fellowship had more prestige than a Ph.D. So I never became a 'doctor' until my first honorary degree, in 1946 from the University of Louvain.

The Cavendish was becoming more and more devoted to nuclear physics. Later in this book I shall describe the rivalry that grew between the nuclear physicists—who would hardly admit that other kinds existed, and the others. Rutherford said—or is thought to have said—'There are two kinds of science, physics and stamp collecting'. By this he meant just collection of information, and by physics I am sure he meant nuclear physics. In spite of my experience with Bragg, I was a whole-hearted nuclear physicist. The one theorist who was not was A. H. Wilson, who was introduced to the theory of electrons in metals by Fowler and spotted the true origin of the difference between metals and insulators. I first heard of this when Fowler was explaining it to Charles Ellis, one of Rutherford's closest collaborators, who said 'very interesting' in a tone which implied that he was not interested at all. Neither was I—though later this work proved to be the foundation of much that I did from 1933 onwards; also, and much more important, it provided the theoretical foundation for the transistor and silicon technology. But at that time I was as nuclear as anyone else. I had a lot of fun with Ellis in examining his experimental results on the β-decay of radioactive nuclei and asking if the facts were consistent with the conservation of energy. We were within an inch of predicting the existence of a new particle, later called the neutrino, but alas we just didn't have the courage to do it.

International conferences were a feature of life. Just before I took up my Cambridge job there was one on atomic physics in Leipzig where I went with Ruth; afterwards we went on to Dresden and I am fortunate to have seen that beautiful city before it was destroyed in 1945. I also remember a conference on the same subject in Rome in 1932 where I was invited to give the opening paper. It was in term time; I could only go away for a week and travelled in a sleeper on the Rome express through the enchanting scenery of Italy in autumn. It was organised by the greatest of Italian physicists, Enrico Fermi, who left Italy shortly before the war and led the team that built the first nuclear reactor in Chicago. The *eminence grise* was however Marconi, very much in favour with the fascist regime and intimate with Mussolini. He arranged that we should meet him; the whole conference was shepherded to the Palazzo Venezia and there was the Duce, on a platform in the corner

Figure 9. International Conference on Atomic Physics in Rome, 1932.
Left to right, front two rows
P. Debye, O.W. Richardson, R.A. Millikan, A.H. Crompton, Mme Curie, G. Marconi, N. Bohr, F.W. Aston, Bothe, C.D. Ellis, E. Fermi, A. Sommerfeld
The author is at the back on the left, exchanging a joke with P.M.S. Blacknett.

of a large room. He addressed us in French, said that Marconi had explained to him something of what we were discussing and wished us success. I think we were all impressed; I remember that Bohr was. A veritable renaissance patron of the sciences, we thought him, and the man who made the trains run to time. Or so we believed.

Fowler had written a monumental book on statistical mechanics which was published by the Cambridge University Press, the same firm that had issued my little book. Fowler then deserted them for the Oxford Press, to set up the far more saleable *International Series of Monographs on Physics*. He and Peter Kapitza (of whom more later) were the editors. They attracted the stars of the physics world, including Dirac, whose book *The Principles of Quantum Mechanics* went into innumerable editions, Gamow on the atomic nucleus, Van Vleck on magnetism and Frenkel of Leningrad on liquids. Fowler suggested to me that I should write for him on atomic collisions, and I found a collaborator, Harrie Massey, an Australian experimental physicist working in this field. Later he became Sir Harrie and had a distinguished career as professor in London and Secretary of the Royal Society. The book appeared in 1933, and went into several later editions, for which Massey was responsible, as my interests had turned in other directions. Actually I wrote, with co-workers, three more books for the series and, after Fowler's death in 1945, I became an editor. Friendship with successive people at the Oxford University Press has always been a pleasure.

During these years Ruth and I got to know Dirac quite well. I believe we were the first people ever to take him to a theatre. He hated anything that was not crystal clear and logical, and did not like the theatre very much. In Göttingen they used to say that he believed that there is no God and Dirac is his prophet. In a letter to my parents (8.3.31) I wrote:

> Cambridge
> 8.3.31
>
> Icy weather here—as everywhere else. I went down to London yesterday in Dirac's car—very cold. Dirac ran—very gently—into the back of a lorry and smashed a headlamp.
>
> Dirac is rather like one's idea of Gandhi. He is quite indifferent to cold, discomfort, food, etc. We had him to supper here when we got back from the Royal Society in London. It was quite a nice little supper but I am sure he would not have minded if we had only given him porridge. He goes to Copenhagen by the North Sea route because he thinks he ought to cure himself of being sea sick. He is quite incapable of pretending to think anything that he

did not really think. In the age of Galileo he would have been a very contented martyr.

Dirac was very popular in Russia; I am sure he did not notice the discomforts of life there. He was invited once to go climbing with some Russians on the Soviet–Chinese border and I remember how he went, dressed in the tidy black suit he always wore, to practice by climbing trees on the Gog-Magog hills outside Cambridge. Later he married the widowed sister of the Hungarian–American physicist Eugene Wigner and they had two daughters. They were neighbours of ours in Cambridge after 1954, and Ruth and Mrs Dirac were close friends.

Towards the end of my second year in Cambridge, an invitation came to be Professor of Theoretical Physics in Bristol. Arthur Tyndall, Professor of Physics there, had made friends with a member of the Wills family, talked to him about physics and got him to finance the building of an enormous laboratory for the subject, quite out of scale with anything else in the recently founded university. Then he had to staff

Figure 10.   Professor Arthur Tyndall FRS.
(from a portrait in Bristol University).

it, and obtained more money from the family, from the Rockefeller Foundation and from elsewhere. He had the right idea, believing that a flourishing research school needed a Professor of Theoretical Physics and secured funds for that too. It was the Melville Wills chair. The first man to be appointed was J. E. Lennard-Jones, but the Cambridge chemists were becoming interested in theory and in the summer of 1932 he left to take up the new Chair of Theoretical Chemistry there. Tyndall had to find someone else. Alan Wilson and myself appeared to him as the only two people with the right qualifications not already holding a chair, and Tyndall asked me. Rutherford gave me a good testimonial as follows:

October 13th 1932

Dear Tyndall

I received your letter asking my opinion of Mott as a possible Professor of Theoretical Physics in your University.

I have come closely in contact with Mott, and he is giving a course of lectures in the Cavendish Laboratory as well as for the Mathematical Board. I can unreservedly recommend him as an admirable candidate. He is a highly trained mathematician and in addition has an original mind; this is clearly shown in his papers, while his gifts of clear exposition are obvious in the book he has published. I do not know of any of the younger men I could recommend with more confience than Mott. I am sure you would find him a valuable addition to your Department and the University.

Yours sincerely
Rutherford

I was very torn; Cambridge was so much the centre of my subject. I went down to see Bristol and said no. Then I went down again, stayed in a house in the Paragon, an exquisite Georgian terrace with a view over the Somerset Hills, and on a beautiful July morning decided to think again, if I could have another year in Cambridge first, which was agreed. Fowler as always was helpful; 'You want to do research, and in Bristol, where you will have few other duties, you will have much better opportunities than there would be in Cambridge. The College will make demands on you—college council, tutorship, the committee life'.

So I decided to go, and began in 1933 at the age of 28, the year when Hitler came to power and in which our manpower in physics in England was greatly doubled through the influx of Germans of Jewish extraction. It was 21 years before I left. I have never regretted the decision.

Before I leave Cambridge and enter quite a different branch of physics, I will try to give some account of how a theorist went about his job in the Cavendish. I had for example to find a problem for Harold Taylor. Now the atomic nucleus of a radioactive material gives out α-particles, which are fast-moving helium nuclei, β-particles which are fast-moving electrons and γ-rays, which are exactly the same as X-rays, namely electromagnetic radiations of very short wavelength. Some of the γ-rays get out, but a good many are absorbed within the atom by the electrons which are outside the nucleus. We knew all about the outer electrons; Dirac's relativistic wave equation had shown us how to deal with them; but we knew very little about the nucleus. We supposed that its internal structure was capable of existing in various 'stationary states', and that the γ-rays were emitted by just the same mechanism as X-rays were within an atom. We therefore made several assumptions about those stationary states, and calculated what we called the 'internal conversion coefficient' for each. From this, by comparing with experimental data, we could find out something about the nucleus. This was the calculation, written into a thesis discussing the whole subject, for which Harold Taylor got his Ph.D.

I need hardly say that in this period we had no idea of the fateful results of our activities, particularly of the discovery of the neutron. That did not come till 1939 when fission was observed, and with it the possibility of a bomb. As Mallory climbed Everest 'because it was there', so did Rutherford and his followers uncover the atomic nucleus.

CHAPTER NINE
Bristol 1933–39

I went to Bristol full of the ideas that I had learned from Bohr, believing that here was the opportunity to create a school where the experimental people and my theorists would work hand in hand. I didn't want a separate department, as did my predecessor; all my ambitions were on working together. In Cambridge I had been a nuclear physicist, and this was because that was the main interest of the Cavendish. In Bristol I wanted to see what people were already doing, and use my understanding of quantum mechanics to help. Even in my last year at Cambridge, when I knew I was going to Bristol, I began to do this.

As my group in Bristol played a central role in the development of what we now call solid state physics, it is interesting to note that the foundations were laid, some years before I came, through the initiative of no less a person than F. A. Lindemann, later Lord Cherwell, Winston Churchill's friend and scientific advisor. He was then a member of the government's Advisory Council for Scientific and Industrial Research. In 1930 he and others were concerned at this country's neglect of fundamental work on the behaviour of electrons in metals, and persuaded the Council that something ought to be done about it. My predecessor Lennard-Jones was told that he would receive the necessary funds if he would undertake to devote some of his time to the subject. This was hardly to be refused, and as a result Harry Jones, a graduate of Leeds and post-graduate student of Fowler's, was appointed as senior research assistant in the laboratory, with the task of studying what had been done and where to go from there.

What had already been done was a great deal. The work of such people as Sommerfeld, Bloch and Peierls in Germany had solved many of the problems which could not even be attacked before the advent of quantum mechanics. In 1930 Alan Wilson explained the origin of the difference between metals and insulators, and Hans Bethe (who will appear later in this book) and Sommerfeld had written a very full account of what had been done before 1933 in the *Handbuch der Physik*. All this was, in the main, about metals in general and hardly

Figure 11. H.H. Wills Physics Laboratory, University of Bristol, as it was before the war of 1939–45.

distinguished between one metal and another—and so was not yet of interest to people in the metallurgical industry. Also the most drastic 'approximations' were made in the theory. With billions of electrons hitting each other all the time, it was possible to get a long way by neglecting this altogether and treating each electron as if it were uninfluenced by the others. I thought it our job in Bristol to extend the theories further beyond both these limitations.

However, Lennard-Jones' heart was not so much in this subject, as in some branches of theoretical chemistry and in 1932 he left Bristol to take up the chair of that subject in Cambridge. Harry Jones then felt out on a limb, and wondered whether he ought to find another job, and was, I am sure, relieved to find after my appointment that I was intensely interested in the work he had done. He had been in contact with the leading metallurgist in Oxford, William Hume-Rothery who, with the minimum of support from the University and although completely deaf, achieved an international reputation and the fellowship of the Royal Society. He had noticed that the alloys of copper, such as brass (copper–zinc) or bronze (copper–tin) achieved the same crystal structure when the ratio of the total number of atoms to the number of outer electrons in the atoms reached a given value. By the number of outer electrons was meant the number that, in chemistry, could take part in chemical bonds; it would be one for copper, two for zinc or four for tin. Here was something new, which the chemists would never have guessed. Jones, by supposing that all these electrons really were running about freely in the metal, showed that quantum mechanics could explain what we called the Hume-Rothery rule.

It was a revelation to me that quantum mechanics could penetrate into the business of the metals industry. Jones published his ideas, and we set to work together to write our book, *Theory of the Properties of Metals and Alloys*. Wilson's *Theory of Metals* appeared a little later. Wilson—in the tradition of the German theorists who began it—set out to do the mathematics properly. We based our book very much on Bethe's *Handbuch* article, and tried to extend its influence by sorting out the differences between real materials, and making approximations and using intuition whenever we liked.

Our book was a success, but not everyone shared our point of view. Even after the war, Professor Smolochowski in the USA records in *The Beginnings of Solid State Physics*[3] how

'Mott ... and his colleagues horrified us by their 'simple' visualizable and seemingly uncomplicated models and mathematics ... In the 1940s and early 1950s, Mott's almost eclectic point of view was often criticized as being unsound and contrary to the need of a unified all-embracing solid state theory. I remember very well the heated arguments starting with the metals conference in Bristol in 1934 and continuing later at nearly all such gatherings including the famous Varenna School in 1952. On those occasions Mott stood up, or preferably walked in his stilt-like manner, and shaking his head slowly from side to side, explained patiently his model, comparing it with experiment. To many of us this was not *gründlich* enough. Slowly, slowly the influence of Mott and of his students and the success of their approach became so evident that no-one questioned seriously their value'.

Another member of the Department was Herbert Skinner, an

experimentalist, whose work on X-ray emission from metals led us to an understanding of why and when it was legitimate to neglect the interaction between electrons in metals[4].

Thus the department consisted of people (Jones, Skinner, Sucksmith, Potter) who could be and were absorbed into a group working on the properties of metals. I could suggest what experiments they should do and try to see what their results meant in terms of quantum mechanics. In addition to mine, there was another group under Cecil Powell, also from Cambridge, who worked on cosmic rays. These are very fast electrons and ionized atoms which reach the earth from outer space, and which hit atoms in the atmosphere of any matter placed in their way, breaking them up—just the same process as the 'atom smashing' carried out by Rutherford with alpha particles from radium. Powell developed a clear and effective method of seeing what happened by looking for 'events' recorded in the emulsion of a photographic plate. After the war he sent his plates into the upper atmosphere in

Figure 12. Research group at Bristol, about 1935.
Left to right
Back row: R.W. Gurney, W.R. Harper, H.C.B. Lovell, N. Thompson.
Middle row: K. Fuchs, H. Jones, E.T.S. Appleyard, W.G. Baber, C.F. Powell, W. Burrough.
Front row: W. Suchsmith, H.W.B. Skinner, Gladys McKenzie (Mrs Harper), L.C. Jackson, N.F. Mott, A.M. Tyndall, S.H. Piper, I. Williams, H.H. Potter, W. Heitler

balloons. He had great success and was awarded the Nobel Prize for physics in 1952 for the discovery of a new particle (the 'pi meson').

The reason for my not throwing my lot in with Powell and applying quantum mechanics to his problems was that I did not know how to do it. We did not then have the knowledge to apply quantum mechanics to the structure of the atomic nucleus; something new was needed, and indeed the pi-meson turned out itself to be a kind of glue which stuck neutrons and protons together in the nucleus. Of course there were things to be done in the nuclear field. Gamow's work on radioactive decay, and mine in Cambridge on internal conversion of gamma rays, dealt really with what happened just outside the nucleus. Inside, all was unknown, until the big machines and the most talented theorists began to unravel it, well into the post-war period. I chose to work in areas where I knew quantum mechanics would work and where there was even more work to be done. Also, I was attracted by the possible interaction with the world of industry.

As I have said, 1933 was the year that Hitler came to power and of the exodus of most German physicists of Jewish descent. The number of theoretical physicists in England must have doubled through the influx. Lindemann, I believe, took his Rolls-Royce to Germany and collected some of the best physicists for Oxford, completely reviving Oxford's Clarendon Laboratory. In Bristol, we had about six, enormously strengthening the theoretical group. One of them was Klaus Fuchs. He was not Jewish, but a political refugee, with communist sympathies. This was acceptable to us—anyone who was against the Nazis would have been. I remember that he arrived at my office out of the blue; he was staying with Ronald Gunn and his wife, Quakers who worked for Imperial Tobacco, and indeed remained there during the whole of his four years in Bristol. He seemed to know his stuff, so I took him on and he produced some excellent work. He was shy and reserved and I do not remember discussing politics with him. After four years I arranged for him to go to the former leader of the Göttingen theorists, Max Born, by then professor in Edinburgh. Born, in his autobiography[5], writes that I wanted to get rid of him because he was a communist, but that was not so; we had many refugees in Bristol and needed to think about permanent posts for some of them, and we hadn't the resources to provide for all.

Fuchs' subsequent career is part of history. In 1940 he was interned as an enemy alien and sent to Canada. But he was released and returned to Edinburgh in 1941. By this time physicists were very scarce in Britain and most of them were involved with radar. Fuchs went to work with

Peierls at Birmingham University on the early atomic bomb project and was given British nationality in 1942 as a 'national interest' case. I signed the application for this, certifying his loyalty. In 1943 he went as a member of the British team, which joined the US atomic project, and worked there on gaseous diffusion and then in the Los Alamos Laboratory on the 'implosion' mechanism for the plutonium bomb. He passed on copious information to Soviet contacts in both countries. After the war he was head of the theoretical division at Harwell. He came under suspicion, was arrested, convicted and imprisoned in 1950. When I heard the news I felt it to be incredible, but it was only too true. In fact the intelligence people descended on me and grilled me on why I had signed his naturalization certificate—not at all a pleasant experience.

After Fuchs was released he was deported and settled down in East Germany, in Dresden. In 1979 I visited that country, as the guest of their Academy, the Leopoldina. I gave some lectures in Leipzig and Halle; my host said he had invited Klaus Fuchs to come over and meet me, but he had said he would rather not. I am grateful for that. I could not have embraced my long-lost pupil with 'all is forgiven', neither could I have shown any hostility to a man who had done what he thought right and had suffered for it.

In Bristol in the 1930s, we had a branch of the Society for Cultural Relations with the Soviet Union. It met from time to time in a studio in Park Street, which disappeared in 1940 in the first big raid on Bristol, (during which I remember walking home from a meeting, with incendiaries falling in the street). We used to dramatize translations of the Soviet treason trials, by which Stalin appears to have got rid of most of his possible rivals. They were accused of sabotage in the interests of the Germans. I was an enthusiastic member, though I found the story incredible. But my most vivid recollection is of Fuchs in the role of Vishinsky, the prosecutor, accusing the defendents with a cold venom that I would never have suspected from so quiet and retiring a young man.

As was natural in the Nazi period, most of us were politically committed and were friendly to the Soviet Union because they were against the Nazis. But I remember my first doubts about the Left in politics. Some of my friends believed that the National Government of the 1930s (Conservative in essence) should be opposed in all that it did, and since it was proposing some limited rearmament, that should be opposed too. Since this seemed to me to be neither in the interests of our potential ally, Russia, nor in our own, I became suspicious of the Left. On the whole this suspicion has remained.

The opportunity to visit the USSR came in 1934. The Soviet Academy arranged a conference to celebrate the hundredth anniversary of the birth of the great Russian chemist Mendeleef, who first discovered the periodic table of the elements. Why a young theoretical physicist aged 29 should have been invited I never discovered; perhaps because I had helped one of their leading theorists, J. Frenkel, to prepare a book for publication in English by the Oxford University Press. Professor Peter Kapitza, a Russian physicist who had been working with Rutherford for some years, was also going. He used to boast that he was the only Soviet citizen who had his passport endorsed for unlimited journeys in and out of the USSR. In Cambridge he ran the Kapitza Club—an evening meeting where we all talked about the latest research. I knew him intimately and he and his wife invited Ruth and me to share his car in a journey from Bergen, round the Gulf of Finland, to Leningrad. In the event there was not room in the car for us and the luggage and we went on our own by boat; I remember however the impression his shiny Buick and the GB plate made in the empty streets of Leningrad.

When he got there, the Soviet authorities told him he must not return; he was to stay in Russia and build an Institute to do what he was doing in Cambridge. What almost destroyed him was the lack of trust. He was willing enough to serve his country, but not being allowed to go back to Cambridge and wind things up hurt him deeply. Also, in the first year, while the Institute was being prepared, he had nothing to do and the thought of the beautiful science he might be doing in Cambridge was torture. All this is revealed in his letters to his wife (see ref. 6).

Ruth and I took a Soviet boat from London for the five day trip through the Kiel Canal, manned by Nazi storm troopers, to Leningrad. Sidney Webb was on the boat, on his way to write *Soviet Communism, a New Civilisation?*. The conference, a vast crowd of Russians and some twenty foreigners, consisted of orations in Russian about the achievements of Mendeleef and after the first session we absented ourselves, met the Russian physicists and saw the sights, ate a large worker's lunch in a turbine factory and enjoyed that beautiful city in the September sunlight. To me, from England at the height of the depression, Russia appeared as the country without unemployment. At any rate I wanted to believe in it. It was after the 'dekulakization' but before Stalin's purges. 'What about the Kulaks?' I asked a Russian physicist. 'Well, we had to get rid of the half a million rich peasants in the interests of the masses, but now that this has been done there will be nothing more like it, and the future is rosy.' I believed him. And they said the Kulaks had been absorbed in the industries of the north. Ruth was more sceptical.

I have a very vivid memory of returning with Frenkel to his house after a long day totally without food, at the laboratory (the Ioffe Institute). Frenkel said lunch would be at four and I was hungry. When we got there, he went upstairs to see his wife, and came down with a beaming face, saying 'I was mistaken; lunch will be at six, and this gives me two hours to explain my latest theory'. It turned up at 6.30. His son, writing a biography of his father many years later, was delighted by this story.

Other memories of Russia are the park where you could gain a prize by throwing balls at Hitler, Austen Chamberlain, Mussolini and Ramsay MacDonald, and the pride of everyone we met in what the country was doing and the apparent hope for the future. However I remember when we returned via Finland and Stockholm, in the Finnish sleeping car in the Finland station in Leningrad the attendant said with pride as we looked at the jug of drinking water, 'Pure water from Finland'. One did not drink the tap water in Leningrad.

Back home, after the publication in 1936 of 'Mott and Jones' on metals, I turned to semiconductors, with the help of Ronald Gurney who came to us from Manchester at that time. He had worked in Cambridge and Manchester; he approached the Bristol Laboratory to ask if we had a place, and we were able to find him a small salary. Together with Condon he had given the first explanation, based on quantum mechanics, of the alpha-decay of radioactive nuclei. This was the theory usually associated with the name of Gamow, (see Chapter 5) who developed it in greater detail. Gurney had a remarkable talent for understanding theoretical physics in terms of pictures of atoms, waves and so on and with hardly any mathematics at all. In 1940, as he was over 40 (the age for compulsory military service), he went with his wife to Sweden, and thence through Russia and Japan to the United States. He stayed there after the war, but because of the sympathy that he and his wife showed for liberal causes in the McCarthy era he was interrogated by the Federal Bureau of Investigation. I do not know details, but believe that this may have contributed to his premature death through a heart attack.

A semiconductor is not a metal; it contains no free electrons but it does contain a few that are very weakly held in position, so that as the material is warmed up the electrons become free. Alan Wilson had explained this in 1932, but there was plenty of work still to do. Some people may remember the wireless receivers of the twenties, in which one had to press a wire against a galena crystal and find a spot which 'rectified' the current, that is, allowed it to pass only in one direction. I published a theory which stood the test of time on that and a good many other papers. Ronald Gurney and I set to work on a book, *Electronic Processes in Ionic Crystals*, which came out in 1940. This was inspired by the experimental work of Robert Pohl, in Göttingen. I had met him at one of the conferences we organized in Bristol, and as I consider him to be one of the true fathers of solid state physics, I should say something about him. He worked with crystals of sodium chloride (common salt) or potassium chloride. These have the chemical formulae NaCl and KCl respectively, which means that sodium chloride contains exactly as much sodium as it does of chlorine. The crystals are transparent, but if they are heated in the vapour of the metal, they acquire a deep colour, blue in the case of potassium chloride. This must be caused by a slight excess of the metal in the material. So in the crystal were colour centres (Farbzentren in German, or *F*-centres as we called them). We wondered about the nature of these. The answer became clear after lengthy discussions at a conference I organized in Bristol in 1937, which Pohl attended. They were places where the anion (e.g. the negative ion of chlorine) was absent, and an electron was present instead.

Why was this so interesting that Gurney and I could write a book about it which, after the war, was quite influential? The answer is that, in the science of solids, traces of impurity can often profoundly change the properties of the material. They can make it conduct millions of times better, make it stronger, or give it colour. A German scientist, Smekal, had written a rather verbose article saying that this was so, but Pohl was the first to study the phenomenon in detail for one rather simple kind of material. This study led to a great improvement in our understanding of semiconductors, which became vastly important after the war.

Pohl himself—like Rutherford—said that he despised theory; *Papierphysik* (paper physics) he called it. And yet the succession of short research papers which he put out from Göttingen were so clear that the correct explanation leapt to the eye. He also said, in an interview in the 1970s (see ref. 3) that the men in the Göttingen school, who invented quantum mechanics and, under Professor James Franck, did so much on the experimental side to prove it, tended to despise his work on 'dirty crystals'. He said that, when my Bristol school and that of Fred Seitz in America took an interest in what he did, this was a welcome support.

But by then, of course, the Göttingen school of atomic physicists, mostly Jewish, had dispersed to England and America.

In 1938 I had my first invitation to America, to teach in a summer school in the University of Pittsburgh. Crossing the Atlantic in the *Queen Mary* and taking the train to Pittsburgh I found a university not of the first rank. The astonishing 'cathedral of learning, a gothic skyscraper looking as if it were made of rubber and pulled out to its height of forty stories' was only half in use. The head of the university, hired I believe from Hollywood as fund-raiser, expressed the hope that 'amid the smoky hills of Pittsburgh a new and better Athens would arise'. The morale of the staff was low and for my first weeks there, astonishingly in the country of often overwhelming hospitality, I was rather lonely. Later I met people involved in the social problems of the city, and in its industry. At the end of the school, Ruth joined me and we travelled across the continent by train, with stops in the Rocky Mountains, to Vancouver where a cousin of Ruth's and her husband were living. Then to Seattle, where Ruth had an uncle and thence to Berkeley where another cousin put us up for a fortnight and showed us the country. I remember my surprise that the painters who were decorating their house all turned up in their own cars. Then back via Los Angeles, the Grand Canyon and Wisconsin, giving lectures here and there, to New York. During this journey Neville Chamberlain made his fateful journey to Germany and war seemed imminent.'Surely you are not going back to *England*?' people asked us. By the time we got to New York, Chamberlain had returned with 'peace with honour'. Our last engagement was in Boston, a hurricane had washed away part of the railroad, and we took a plane—Ruth's first flight. The wind had flattened the innumerable bill-boards that disfigured the countryside, making us think of the devastation in our own country that we had so narrowly missed and expected later on. We took the boat at New York somewhat depressed.

In America I had several discussions with Frederic Seitz, perhaps the father of solid state physcis in that country, whose massive book, *Modern Theory of Solids*, covered much the same ground as mine. I remember particularly explaining to him Landau's conception of a polaron in ionic solids, as an electron which 'dug its own (potential) hole', a concept that he had not previously understood.

One of the last pieces of work that I undertook before the war was my theory with Gurney of the photographic latent image. How did light, absorbed all over a grain of silver bromide in a photographic plate, succeed in producing a speck of silver somewhere on the surface?

This was called the 'latent image'. Our understanding of the properties of single crystals came from fitting a theory to Pohl's work, and this enabled us to see how it happened. Our success gave me my first intimate contact with industry, namely with the Kodak research laboratories in Wealdstone and after the war, in America. In 1940 I received the Harker and Driffield medal from the Royal Photographic Society for this work.

I have often wondered how much help this was to the industry. One of the Kodak people remarked that if photography had had to wait for the Gurney–Mott theory, it would have put things back for fifty years! This is true; like so much of technology, the photographic plate and films were developed long before anyone understood *why* they worked. But my friends in the industry assured me that, with the theory, work on adapting photographic materials to new conditions and improving them in many ways was greatly helped. I hope so—but it is impossible to quantify this.

In 1936, at the age of 31, I was elected into the Fellowship of the Royal Society. This was younger than usual, but nowhere near a record; Paul Dirac was elected when he was 28. I need hardly say that I was pleased, for the Fellowship is greatly prized. In later life, however, I have often wondered if it is healthy for the scientific community to make such a sharp division between those who have achieved it and those who have not. On the borderline there must be a lot of luck, and those who just miss it can find it hard.

After this outline of my professional career up till the war, I must try to say what I think I achieved, in perhaps in my most productive period before 1965. After the war, I found myself to be (apart from Dirac) the outstanding theoretical physicist in the UK, and was offered chairs at Cambridge, Oxford, London and appointed President of the International Union of Physics and so on. In fact, there were very few rivals in my field. If I ask myself what this reputation was based on, it must be the researches incorporated in my two books, Mott and Jones (metals) and Mott and Gurney (semiconductors). I feel that nothing we did was difficult, there was no brilliant breakthrough in our understanding of nature, there was nothing of Nobel Prize calibre as in the work of a Heisenberg or a Dirac. On the other hand my contribution was unique, perhaps, in taking a new subject that was outside the mainstream of pure physics but of some interest to industry, in choosing fruitful problems, explaining them clearly and inspiring a research group. I feel so often in scientific research that, had I or someone else not become successfully involved in a given problem, another worker would have done it a year later. But I don't feel like that, for instance, about our theory about the photographic latent image. I don't think anyone would have thought of the problem as one in which quantum mechanics had anything to say, at any rate not till many years later.

Turning to more personal aspects of life in Bristol, Ruth and I had a flat (4 Caledonia Place) near the Clifton suspension bridge. Laboratory walks in the Cotswolds and Mendips were features of weekends. Ruth did some classics teaching at Badminton School and elsewhere. She was Secretary of the Bristol Classical Association. Later—very near the outbreak of war—we obtained a superb flat looking over the Avon gorge (6 Princes Buildings), from which we could watch ships coming up and down to Bristol docks, and later, on one occasion, daylight bombing of the harbour area.

After the German occupation of the Sudetenland there was a movement to rescue children of whole or partial Jewish descent from Czechoslovakia. We put our names down for two, Lilly and Ilse Spielmann, aged 16 and 12, daughters of a Jewish musician and his non-Jewish wife. They duly arrived a few months before the war. Ilse went to an independent school in Bristol. With the outbreak of war and after the fall of France in 1940 Bristol became a prohibited area for 'aliens' over a certain age and the difficulties of finding a home for Lilly were immense. Ilse, under the age of 16, was able to stay with us. My sister and her husband, Dr. Arthur Fitch, who had a boarding school in a remote Yorkshire village, were of the greatest help and came to the rescue.

Eventually after the war, both girls married and now live in England. Their father had perished—we never learned how. The mother survived.

In the last year before September 1939, as war seemed certain, I remember feeling it urgent to complete and publish my book with Gurney; war with the prospect of bombing, many of us believed, would be even more disruptive of the normal life of the country than it actually was. The book appeared in 1940.

In July 1939 no-one I knew believed that war could be avoided. None the less, in July I attended a conference on magnetism at Strasbourg, organized (ironically) by the Committee on Intellectual Co-operation of the League of Nations. My most vivid memory, apart from a trip into the Vosges with the future Nobel prizewinner Louis Néel, was of the speeches at the banquet, the Mayor of the city saying that now all France could do was to prepare its defences and the Germans expressing utterly incredulity that war could occur.

CHAPTER TEN

The War 1939–45

As the war approached I did not make any attempt to involve myself in military research, probably because I did not know what was going on and doubted if I had anything to offer. I think we all expected German bombing to be more totally devastating than it turned out to be and my anxiety was to finish certain researches and give them to the world before the final catastrophe. But a month or so before war actually broke out, John Cockcroft, who was doubtless on committees concerned with defence, approached some ex-Cavendish physicists including myself, telling us 'If war breaks out, most of you will get involved in military research. It would be most helpful if you first got some practical experience of what you will work on. I want you to spend a week or two of the summer on the south coast so as to see the practical side.' I do not remember whether we learned what the highly secret installation was before we got there, but I remember the intense excitement when Herbert Skinner told us 'This is the first line of defence of the country'. It was one of the early radar sets, which could detect an aircraft approaching our coast 50 or 100 miles away, and pass the message to our fighters. At that time they were called RDF (radio-direction finding), and it was put about, and believed in the locality, that these tall aerial masts emitted a ray that would stop aircraft engines. There were stories of car engines being stopped if the car got too near. Under the supervision of air-force personnel, we manned the installation. But when war was declared, the RAF took over; we were sent up the aerial mast to keep watch in case an enemy flew in under the radio beam. Going up that mast was far more terrifying than any air raid that I experienced.

Radar had its origins in the work of Edward Appleton and his colleagues in investigating the ionized layers in the upper atmosphere, which long-wave radio waves cannot penetrate and which make radio transmissions round the earth possible. He sent up radio pulses which were reflected from this layer, and observed how long they took to travel there and back; thus he could determine the height of the ionized layers. However, he found that he also saw aeroplanes. Those interested in the struggle to produce our radar chain in time, should see for instance R.V. Jones's *Most Secret War*[7].

Coming back from the South Coast, I tried to find a job in defence research. I remember making a journey to Dundee to see what was going on there, in a crowded train from which restaurant cars had been removed as a measure of war economy. A research station had been evacuated there from Bawdsey on the East Coast, which was thought to be too vulnerable. I managed to visit other research stations and gun sites. Thinking I would be away most of the time, Ruth took a teaching job at Wells, where St. Brandon's school, to which we had sent Ilse, was evacuated. Our beautiful new flat was kept open by her sister Mary Horder, who had a job in Bristol, and was filled with lodgers, from King's College London and the BBC, which was evacuated to Bristol. So I lived there when I was in Bristol, and Ruth came at week-ends. My first job concerned the G.L. Mark I, the early radar system which was attached to the anti-aircraft guns to enable them to engage the night bombers. It involved matching the aerials to the rest of the circuit; I had done what I could beforehand to read up on the physics of radio. Then came another problem. The sets were mounted on a wire netting mat, which had to be flat. In order to estimate the height of the enemy aircraft, the direct radar wave and the wave reflected from the mat had to interfere, and one made calculations of how they would do so. The wire net was held up by posts stuck in the ground, and to get to the set one crawled under the net and through a hole, and thence into the cabin. But how much error would this hole produce? I worked all this out—was it worthwhile covering it up with another piece of wire netting? I think the error seemed significant, but gun laying by these early sets was so inaccurate anyhow that it could hardly have made much difference.

At the time of the fall of France and the evacuation of the army from Dunkirk, I was still based in Bristol doing some jobs for various establishments. I saw a daylight raid on the harbour from the roof of our flat, and some bombers shot down by our fighters. It was the time of Churchill's great speeches and of a certain exaltation in the public mood; I remember a trip to the Isle of Wight to see a radar station, and wondering if I should be trapped there by an invasion. We spent night raids sometimes on the roof and sometimes in the cellar of the building housing our flat, which belonged to a very old lady of the house of Harvey, whose rows of Bristol milk and cream sherry from her firm were beyond an iron grating which was not opened for us. With the fall of France, our six refugee physicists who had not had time to obtain British nationality were all interned and sent to the Isle of Man. We filled in forms to get them out again, starting with Walter Heitler, the most distinguished and ending with a young student. In the Home Office they must have turned the pile upside down because the youngest came out first, and Heitler, extremely disgruntled, the last. Before the end of the war he was invited to a position in Dublin and later became professor of theoretical physics in Zurich.

In January 1941 Elizabeth, our first child, was born in a nursing home in Wells. The question was—where should Ruth and the baby go—and eventually they went to her widowed mother's house near Hayward's Heath in Sussex, and stayed till the war ended. Except during the period of the flying bombs life in Sussex continued very much as in peace time, even with a maid. In February 1943 our second child, Alice, was born in a hospital in Lewes.

From the end of 1940 I became more and more involved in the operations of Anti-aircraft Command, and eventually was appointed to succeed Blackett as scientific advisor to the commanding officer, General Sir Frederick Pile. I got lodgings at Stanmore, and after some time an office to myself at Bentley Priory, their headquarters. This experience gave me a deep respect for the professional soldier. When the night attacks on London had begun, there was no radar yet for the guns and the scientists told General Pile that it would be useless to waste ammunition, as the chance of hitting anything was minimal. But his answer was 'There is the enemy; here are my guns; we open fire'. And how right he was! The noise and the feeling that we were hitting back gave real comfort to Londoners.

My colleagues and I were constantly preoccupied with the chance of bringing down an enemy aircraft because of the advice we had to give both to our own guns and to Bomber Command. All our gunners could do was to fire into the cubic mile that the aircraft would be in when the shell burst, if it went on flying in a straight line. We did not know how many shell fragments would have to hit a bomber to bring it down—we saw our own bombers which had come safely back with many holes in them. Something could be gathered from studying the rate of success of our own guns in the night attacks at the beginning of 1941, when radar was installed. I went round to look at some gun sites—Plymouth, Portsmouth, London, Birmingham and others. Whether I was told to do this or did it on my own initiative I can't remember; I was allowed a lot of freedom, and usually found myself welcome. A striking fact, which I wanted to understand, was the much greater success of the coastal sites such as Plymouth compared with London or Birmingham;

the former claimed that an aircraft was shot down for every 2000 rounds fired, contrasting with 20,000 for inland sites. I wondered if the aircraft flew in straight lines over the coast and embarked on a zig-zag course inland, but the radar tracks revealed no such thing. At last I thought of enquiring where the aircraft had fallen—and found that they had all been seen to fall into the sea, instead of half of them, as would have been expected. A policeman or a clergyman or an air raid warden had seen a ball of fire falling towards the sea, had reported it and the claim that an aircraft had been shot down was accepted. A possible explanation was that one shell in 2000 did not detonate, but the explosive ignited and dropped in flames to the sea. 20,000 was, I suppose the true figure for rounds per aircraft destroyed, and that was less than one in each raid.

The over-estimate of enemy aircraft destroyed both by fighters and guns, came to be known in operational research circles as the 'lie factor'. When I told my General about this, he looked grave but not surprised and said 'You must not mention this till the war is over'. But it seemed to get around—not, I believe, through me.

The scientific problem of why the shell fragments had the size observed and the operational problem of what is the optimum size for various purposes, kept on recurring. German shells, using steel with a higher carbon content, gave smaller fragments. I wrote a paper on the scientific problem, which was published after the war, and my memoranda written during the war still seem to survive in American ordnance laboratories and have been used. On the problem of the optimum size I worked out that the shells of Bofors guns would be effective in cutting barbed wire, and I remember a day when some of us went down to Salisbury Plain and the idea was tried out successfully. But whether it was used, I never heard.

It was a privilege to work with some of the men at AA Command. I admired their calm and matter-of-fact courage, though perhaps I did not always agree with their opinions. I remember a Colonel Fremantle, who came in freshly shaved one morning after a night over Berlin to see what AA gunfire looked like from the receiving end. A letter to my mother dated March 1942 says:

We had a meeting at Christchurch at 10 o'clock and Fremantle and I left Stanmore at 7 in the morning to be driven down to Christchurch by a girl of the ATS who covered the 100 miles in 2½ hours. Such empty roads—Sunday morning early; in wartime the country is lovely. Fremantle is a tough guy, he starts off with no breakfast and talks all the way down about the relation between the romantic movement in art and the present troubles of the world. He seems to think that it is all because Beethoven became deaf and took himself so seriously that Hitler arose and took *himself* so seriously.

My main concern at Anti-Aircraft Command however was with searchlights. They were, for administrative reasons, clustered three on a site and as a result the operators simply dazzled each other. I visited many sites, often in the lovely country of North Norfolk, and did some quite serious work on how they ought to be deployed. But the man in charge, a Major Smith, a former black and tan in Ireland, did not listen. I could not help rejoicing when he went away sick. A rather indiscreet letter to my parents tells what happened next (17 May 1942):

> We have a new man in charge of searchlights, a young man, and *quite* a different kettle of fish from old Smith who was the ruin of them before. He has been in A.A. Command under Smith in charge of searchlight training, and we never knew what he thought while Smith was there. Yesterday we had a meeting, the C. in C. and his Major-General, this man, the Colonel in charge of 'Ops' and I. Well, there was the Chief explaining to 'Ops' all the points I've been telling him for months, and which he used stoutly to deny, and there was our new Searchlight man telling him even more fundamental things ... *We* get no credit for this, nobody comes to us and says 'sorry old boy, you were right all along ...' but at least I feel that much patient educative work has borne fruit and that we can leave searchlights in peace (I mean we shall feel in peace about them).
>
> Sometimes I wonder if they could have found out how to use them without our help; but in view of past history I don't think so.'

But all this was after the period of night bombing and I doubt if my work on searchlights had the slightest effect on the war except possibly against flying bombs; moreover radar must rapidly have made them obsolete.

Meanwhile an army operational research group was growing up in Petersham near Richmond and it was decided that General Pile no longer needed a scientific advisor but could obtain all the help he wanted from there. With regret I left Stanmore and got lodgings at Richmond. There was a proposal for an operational research group on tanks, though it was not clear what it could do. Operational research had had success in forms of warfare where a similar action is repeated: protection of a convoy, attack by and defence of bombers, finding the target in a night raid. The enthusiasts, John Cockcroft and also Charles Ellis who was now at the War Office, wanted to try it for everything. I was asked to find out what could be done for tanks. I should have realized that with no combat experience it was a hopeless quest. I got books on tanks, visited a training camp, drove a tank and fired its guns, and totally failed to come up with anything. After this I was frustrated and unhappy at Petersham—not helped by my dislike of the man in charge and probably his for me.

About this time I was awarded the Hughes medal of the Royal Society (for my work before the war), and sat on the Sectional Committee which chooses new Fellows. It seemed incongruous to continue with these matters at such a time.

In May 1943, to my relief, I was offered a new job. A letter to my parents records it as follows:

Thursday May 19, 1943

I went over the other day to the (evacuated) research and development department of Woolwich Arsenal (now near Sevenoaks), and called in to see Lennard-Jones, my predecessor at Bristol, who has been appointed Chief Superintendent of Armaments Research there. Woolwich research has the reputation of being an Augean stable, and it was hoped that L.J. could clear it up. He has the reputation of being a good organiser, though I don't think that he is a very good theoretical physicist. He holds the chair of 'theoretical chemistry' at Cambridge. Lennard-Jones, completely to my surprise, offered me a job as superintendent of all the mathematical research on armaments carried out at Woolwich—on a par with the supt. of explosives research, metallurgical research, etc. The job is not yet created but L.J. said he had already approached the relevant civil service high-ups and had gained their assent.

I'm pretty sure that the thing to do is to accept and have already told L.J. so. The work in Army Operations Research Group is probably more interesting, and through our contacts with the War Office and Bomber Command, I get now and then the chance to influence things in some small way or other, perhaps more directly than I would at Woolwich. On the other hand my relations with the man who runs the group have not been a bit happy, and you know how that cramps one's style. The responsibility at Woolwich would be much wider—quite a considerable proportion of the mathematical talent of the country would be under my control, and of course I would concern myself with the experimental work too. I know a lot about it already, and the insight into the user end gained at AORG would be invaluable. Also I am anxious to try my hand at some job that involves some organising. My ambition at the moment for after the war is to start, either at Bristol or elsewhere, an institute for Industrial Theoretical Physics, where we would take scientists from industry and give them courses on the mathematical side (the request here comes from them) and try to do their theoretical research problems. And what better introduction to that than this Woolwich job?

Disadvantages: to leave my pleasant digs in Richmond and the small circle of friends that I have and go to this wretched suburb in Kent [Fort Halstead, near Sevenoaks]; rather further from Two Oaks [where my family was] too. It may be that I shall find it best to shift the mathematicians into London.

Also L.J. is a man about whom opinions are very mixed. Personally I have always liked him.

Well, the thing is not fixed up yet, but L.J. has got the consent both of his boss and my present employers, so I suppose it will be. I shall then enter the hierarchy and have initials SMRA (Supt. Math. Research in Armaments).

My appointment went slowly through the proper channels and it was 1 September 1943 before I took up the job. The parting words at my interview were 'And when, Professor Mott, would you be able to take up this position?' 'In three weeks from the day that you make known your decision.' (I wanted a holiday before I began.) 'Excellent! We do not usually move so fast in government circles'. I got lodgings in Petts Wood. This was easy; billetting was compulsory and most people, women at any rate, preferred one professor to two girl typists.

Lennard-Jones, in moving Woolwich to Fort Halstead, had introduced academic and industrial people, Sykes, Curtis, myself, present and future Fellows of the Royal Society, over the heads of permanent civil servants. Our tradition was that, if we wrote a report, we put our names to it; theirs was that everything was signed, 'Superintendent, Armaments Research Dept.' There was some unhappiness which I tried to assuage. Then I had to go to Cambridge and move a group, which was doing very little good there, to Fort Halstead. They were most reluctant. I wrote in a letter:

The A.R.D. job is becoming extremely interesting, and a useful experience, in spite of distasteful tasks. I am gratified to see my young men getting enthusiastic for their work. The long railway journey and life in billets (however kind) remain disadvantages, and cut down a lot the amount of work it is possible to do. We all find that.

How we all hate the administrative side of the civil service. One of my men from Cambridge (married, house and child) is being transferred to London to do an urgent job. He was interviewed 6 weeks ago, but told nothing though I heard privately that the transfer was agreed by all the persons concerned within a day or two. When I was in Cambridge he had still heard nothing, so I rang up his new employer, and was told that they hoped to get his transfer through the establishment division by Saturday, and that he would then be instructed to begin on Jan. 1st. Imagine the man's feelings, told to move at such short notice, after being kept waiting for six weeks during which time he might have been looking for a house!

A man will willingly make sacrifices to do an urgent job; but when treated like this he loses any enthusiasm or feeling of urgency. It happens again and again.

I'm only now realising that, during the 4 months that they took to get my appointment through, at least 12 young men were doing very little through having no-one to look after them.

The jobs we did at Fort Halstead included a theoretical investigation of the effect of adding aluminium powder to explosives, and research on the mechanism by which armour-piercing projectiles, including 'shaped charges', penetrate the armour of tanks. With the end of the war so near, none of this can have had any effect on actual hostilities. But it was more relevant to the real war when we were shown the intelligence reports of the V2 rocket and asked to estimate its range. We got it right—but whether our report was ever used I never learned. R.V. Jones told me later that he had never heard of it.

Life was somewhat disturbed by the flying bombs; balloons in our neighbourhood were there to bring them down before they got to London—and some did. I saw one explode in the middle of a herd of cows. Another came down at night in the part of the Establishment where explosives were stored, not doing serious damage. We were still searched the next morning to see that we had no matches as we went in to work.

As the allied armies advanced on Paris after the Normandy landings, Lennard-Jones sent for me and told me of the plan to send a group of scientists in with the army to see what research if any the Germans had been doing. This was a rehearsal for the invasion of Germany. I was to lead the party. I cannot describe my joy at this opportunity to be in

France at such a moment. But the war went on, the armies advanced and I heard nothing more. Then Lennard-Jones sent for me again and said 'I'm sorry, Mott, you can't go. The Ministry of Supply, our employer, thinks you should go in officer's uniform as a major, but the War Office says as an officer without tabs, because then they won't have to pay your wife a pension if you are killed. They'll never sort this out before Paris is liberated. We'll have to confine our party to military personnel.'

It was a bitter disappointment. From then on I felt that anything I did would be for the next war or for the Establishment files, and was keen to get out and back to university work. But I did get to Paris in March 1945—a hungry but beautiful city, untouched by the war. It was arranged by Ellis at the War Office; I am not sure what I was supposed to do. I met Professor Sam Goudsmit, Dutch by origin but established in the USA, and asked him what he was doing. He said he couldn't tell me, so I said 'Oh, Manhattan project I suppose', rather to his dismay. He was leading his famous ALSOS mission to find out what he could about the German atomic bomb project.

Throughout the war I knew, of course, that many of my colleagues were in the USA or Canada working on the atomic bomb project; I had been in the early days on the Maud Committee, (a committee to assess the feasibility of the bomb), but had never been asked—to my relief— to join the work in America. It was exceptionally secret; people working on war research were usually allowed to talk to each other on what they were doing, but not so those in the Manhattan Project. I had however a warning that—very much to my surprise—the bomb was about to be tested. I met G. I. Taylor, who told me that he was off to America. I asked him what he was to do, and he said that he couldn't tell me. Now he was not an atomic physicist, but the world expert on such matters as shock waves, so I deduced correctly that he was concerned with the effects of the bomb, and that it was ready for testing. So the news of Hiroshima was no surprise.

After the end of the German war but before the surrender of Japan, the Soviet Academy of Science invited some representative British scientists to celebrate in Moscow to mark the victory over Germany. These included Blackett, Dirac and myself. We arrived at the Royal Society to get our passports, to sleep there on camp beds for the night and go to the airport early next morning. But Blackett and I were handed our passports back with the exit visa to the USSR cancelled. Officials murmured 'Manhattan Project'. Neither of us had been concerned directly, but it was believed that anyone who had done any military research might gossip, and it was essential to convince the Americans of the strength of our security. We were furious—but could do nothing.

I wrote to my parents on 15 June 1945:

> It's difficult to know how much to say about this scandalous affair. Much of the gossip is too much bound up with secret matters, and what we do or don't want the Russians to know, to be repeatable.
>
> But whether or not Mr. Driberg's statement is correct—and nobody pretends that Mr. Churchill's statement has any basis of truth—it appears that any scientist who has any association with war work is liable to be refused permission to visit his colleagues in foreign countries. This is an intolerable situation, and fatal to our hopes of building up European science again.
>
> Blackett, Norrish, Bernal and I have taken certain steps, including a refusal to participate in any further war work until we receive an assurance that it is not the policy of H.M. Govt. to behave in this way. To see Blackett marching out of the Admiralty was a magnificent sight. We hope that other colleagues will join us.
>
> Although the Russians are only too capable of behaving in this way themselves, it is more than deplorable that we should.
>
> The thing of course had its humourous side. The News Chronicle reporter who accosted me as I walked out of the Royal Society on Wed. afternoon with my heavy bag to go back to Bristol—'oh Professor Mott, *do* let me photograph you with your bag'. The press have been onto us like anything—one has much more publicity as a scientist who could not go to Moscow than as a scientist who did.
>
> The thing has been a Churchill–Cherwell decision; I'm afraid the old man, after his magnificent services to the world, is degenerating into an arbitrary dictator, surrounded by the three undesirables, Cherwell, Beaverbrook and Brendan Bracken.
>
> Anthony Eden, by the way, wrote to the President of the Royal Society some weeks ago urging that as many as possible should go. But Eden is ill.

In the last days of the war, I had many offers of post-war jobs. Bragg at Cambridge suggested the Chair of Metallurgy there—a theorist might do quite new things. Later Fowler died and the Chair of Theoretical Physics at Cambridge was pressed on me, as was a similar chair at Oxford. Tyndall at Bristol assured me that, if I returned there, I would succeed him as head of the department. As a theorist I had little competition; Alan Wilson, now an FRS, had decided to make a career in industry. Jobs were chasing me; quite unlike the situation for young scientists today. My thoughts are expressed in the following letter to my mother from Fort Halstead. I feel that my actual course of action was based on this.

I forgot if I told you the latest about the Cambridge metallurgy job, which is that some weeks ago I wrote to Bragg that if I had to decide this summer I would say no. I have heard nothing since.

When one has time to think about it, it is a puzzle to know what to do after the war. It is so patently obvious that what mankind as a whole wants is *not* more applied science, and on the other hand it is equally obvious that what England needs to hold her own industrially *is* applied science, and lots of it. (I speak of applied science rather than pure science because I don't think I'm likely to make a major discovery in pure science, the lines on which I would work wouldn't be in that direction, and anyhow discoveries in pure science turn into applied science with very little delay in these days). I never did think that the world needed more science, but of course research is a thing which compels you with its own fascination, and ambition is a powerful motive too. Only in the war have I felt that the work I did was necessary for some outside cause.

There are various things I could do; I could study biology and biophysics in the hope that I could find some link between it and theoretical physics. This would be a new subject and rather fun; but it might come to nothing amid the vast amount of experimental information available. It would be 2 or 3 years before I got anywhere. The justification is that biology leads to medicine and that more medicine must be a good thing, or at any rate that it can't do any harm.

I don't really think that one can evade the issue like that or take up biology for any other reason than that it seems the most promising line of research. The most obvious line for me, a line that I have talked about and that my colleagues will expect me to do, is to start a school of theoretical research either in Bristol or Cambridge with special leanings towards industry. Of course this began rather before the war, with our work in Bristol on photographic emulsions, but we could make that our more definite aim. We could invite people from industrial laboratories to courses lasting 3 or 6 months where they would be able to learn bits of theoretical physics in which they were interested and attempt little bits of research themselves. I could get some financial help from industry and engage some staff whose definite business it was to go round to industrial laboratories and find out what problems they had susceptible to theoretical research.

In spite of the fact that further technical advance is not the crying need of the world, there are some points of this programme which seem to me good. It would take men out of industrial laboratories and show them new things, make them intellectually more alive. Then it might be possible to arrange that they came in contact with the University outside the laboratory—a little extra-mural influence in the very best sense. Some of these men become powerful in industry and any contacts they can have with humane thought must be all to the good. Then I can't help thinking it would be the basis for a very good teaching school for the students, especially the many who are not so brilliant, from whom the forefronts of knowledge in atomic physics were always a bit too far removed for them to benefit from the presence of a normal school of research.

Cambridge or Bristol—both would have their advantages. Cambridge I suppose ultimately the greater with good students and brilliant colleagues. On the other hand in Bristol a good man with a good programme will have unstinted support from all his colleagues, while in Cambridge that isn't so; there are too many sectional interests especially the college ones which are inimical to research. It is noticeable that provincial professors like their research funds to be administered by their universities, while Cambridge and Oxford men want theirs to be administered by an outside body like the Royal Society. [I found this true, too, after the war.]

Also the only suitable building in Cambridge, as far as I can see, belongs to Lennard-Jones! So unless he decides to stay with the Ministry there would be no accommodation until it is possible to build, and when will that be?

CHAPTER ELEVEN

Bristol 1945–54

We returned to Bristol in 1945, before the end of the war with Japan. Arthur Tyndall, the director of the laboratory and the man who had created it, had been acting Vice-Chancellor and it was doubtless owing to him that the University offered me a beautiful Georgian residence, Stuart House, at the low rent of £70 per annum. It was situated on the campus immediately opposite the physics laboratory. In those days, before the university engineering laboratories were built, both our small garden and those of the university on the Royal Fort site enjoyed the same view of the Somerset hills that had enchanted me when I decided to move to Bristol in 1932. There were many advantages in living on the same site as the laboratory. People dropped in informally. After teaching was over I could take the children into the lecture rooms and let them chalk pictures on the blackboards to their hearts' content. We established warm friendships with both the academic and technical staff of the laboratory, and with many of our colleagues in other faculties too.

On the other hand it was a time when we suffered increasing anxiety about the health of our elder daughter. For a time she was able to attend a normal day school, but before we left Bristol in 1954 she went to a special residential school for the handicapped, where she was happy and cared for by a devoted staff. In Bristol we were fortunate in finding a home help, Peggy Fisk, who with her three-year old son came to live with us as a member of the family. For a time Ruth's sister, Mary, who had a job in Bristol, shared our house too. With this assistance, Ruth was able to teach classical languages both at the university and in local schools, and this meant a great deal to us both.

During our first months in Bristol I remained undecided whether to stay; the Chair of Theoretical Physics at Cambridge was still on offer. Cambridge, with its prestige and ability to choose the most talented students, was always attractive. On the other hand, Bristol gave me the opportunity of doing what I had planned during the last year of the war, with the full support of an ambitious university. This was to create a department of physics where theorists and experimentalists would work

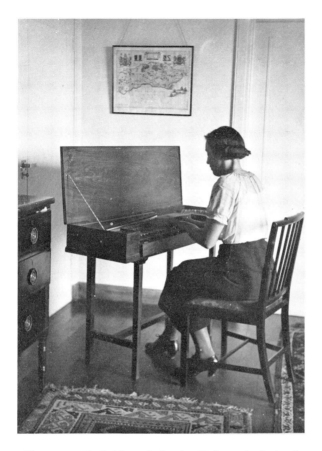

Figure 13. Ruth Mott playing her Dolmetsch clavicord.

together in the closest possible way, and which had extensive contacts with industry. In Bristol people in authority would know what I was trying to do and would want to help; in Cambridge, it seemed to me, there was no-one in authority. Little incidents showed this. I asked Bragg whether I would have a secretary in Cambridge and he said I could share his. In Bristol it went without saying that I would have my own. Nor was the prospect of finding a house in Cambridge and moving to it attractive. Eventually I made up my mind, deciding to stay in Bristol. A letter to my mother, in January 1945, says that various senior people in both universities had commended my decision but goes on

> ...On the other hand we are all rather unhappy about Bragg who
> has a rotten job succeeding Rutherford, in a lab in which there is

a strong anti-Bragg party. Bragg of course is all for metals and contact with industry while the anti-Bragg party wants to go on with the nucleus. I am so very much the obvious man to strengthen Bragg, and I am afraid that he must take it as rather a blow that I'm not coming. I've been feeling most unhappy about it this week as I like the man very much. But if I were leaving Bristol I'd feel worse about them...

The electors at Cambridge then approached Rudolf Peierls, who had ambitions similar to mine for Birmingham and also refused, then Casimir and Kramers from Holland, (see ref. 7), and finally solved the problem by the appoitment of Douglas Hartree.

I look back on my work during these nine years in Bristol as being very satisfying. As I had hoped, we were able to organize summer schools for people in industry, to tell them about new advances in solid state physics, particularly those in the United States. I often meet people who attended those schools, who remember the vigorous discussions between academics and industrial people, and who are grateful for them. One reason why this work was so satisfying was because of the illusions we had. We felt that science had won the war, and that science could win the peace, that British industry would come out on top and poverty would be abolished. Alas, as it turned out, it would have needed more than our summer schools for all this to happen!

When I came back to Bristol, Arthur Tyndall was still head of the Department of Physics. He had promised me that if I returned I would succeed him, as I did in 1948. It did not make very much difference, as we had worked so closely together before and his advice was always available afterwards. He was immensely proud of his laboratory and all it did, and so was I. In everything I did, too, I was helped by our new Vice-Chancellor Sir Philip Morris. (Morris was born in 1901 and took Modern Greats at Oxford. He had been director of education for Kent, and director of Army Education from 1944 to 46 and had held various positions in educational administration before this.) He knew everything, indeed more about my technical staff than I did myself. If I wanted anything, a new post, a change in the regulations, some alterations to our buildings, the request had to go through the proper channels, but Morris would tell me at once whether it would be successful or not.

My first job in the Physics Department was to appoint new staff; Skinner had become director of a division at Harwell, Gurney was in the USA and Harry Jones had gone to Imperial College, where he was appointed to a Chair of Mathematics in 1946, Heitler was in Zurich and

Fröhlich in Liverpool. Of the people I appointed, one, J.W. Mitchell had worked at Fort Halstead and I had known him there. At Bristol he became interested in the theory of the photographic latent image that Gurney and I had proposed before the war. He investigated what happens when light decomposes silver bromide crystals; this is described in Chapter 14. He achieved a Fellowship of the Royal Society and eventually moved to a chair at the University of Virginia. Another, Charles Frank, had been part of R.V. Jones team, (see ref. 7) during the war, and in Bristol had great success in his work on dislocations, also described in Chapter 14. He achieved an FRS and a knighthood, and was awarded a personal chair at Bristol. Dirk Polder, from Holland, was appointed as our senior theorist; he later went back to Philips at Eindhoven. I also had some senior research men. Among them was a thirty-year old Frenchman, Jacques Friedel, whose physics career had been held up during the German occupation. He joined me for a year and stayed for three. Before he left, he married my wife's sister Mary. He became the leader of French solid state physics, and Bristol awarded him the degree of D.Sc., Honoris Causa, in 1975. Their two sons, now both married and with children, are making successful careers in medicine and in physics. Nicolas Cabrera from Spain worked with me on a theory as to why aluminium and stainless steel do not rust. I also published my first paper (in 1949) on 'metal-insulator transitions', an evolving subject which has remained one of my interests (see Chapter 16).

This was the period which saw the invention of the transistor in the Bell Laboratories in America, and with it the beginnings of the microchip revolution. Solid state physics had lagged behind in the UK during the war, but had shot ahead in America. One of our aims in our summer schools was to inform people in industry about what was happening in this field. Also we had contacts with the metallurgical industry through our work on dislocations and corrosion.

During this period I spent less time on personal research than before the war or after 1965. I was intensely interested in what my staff was doing but perhaps less so in doing research myself or having research students. I was also interested, under the influence of Philip Morris, in how the university was run, and in taking my part in running it. Morris had suggested that I should be Dean of the Faculty of Science, a job that would put me at the heart of the administration. I also did a good deal of travelling. I wrote to my parents in 1951

> The thing that takes time and energy is having research students in
> theoretical physics—and it's only sometimes rewarding. To tell you

the truth—at the moment it's very hard to think of research problems that will come out! Next year I'll leave them all to Frank and Polder; I think that will make it possible to be dean of the faculty and keep in touch with my more senior colleagues and do a bit of research myself. I hope so. I'm getting so that I don't really want to get out of being dean . . .

During this period my colleague, Cecil Powell, whose work in cosmic rays had developed strikingly, was awarded a Nobel Prize for the discovery of the pi-meson (see Chapter 9). This was an occasion of great rejoicing in the laboratory. We felt ourselves at the top in both particle physics and solid state—the 'best lab in the country'.

The fact that we thought of ourselves in this way, however, did not mean that we necessarily got the best qualified students; we believed they still favoured Cambridge. For instance, I remember that a brigadier, whom I had known during the war, arrived with his charming daughter who wanted to enter our department to do physics; we showed them round and they were well pleased, but to my chagrin he said as they were going, 'She's in for Newnham; I don't think she'll get a place, but *of course* if she does she'll accept it.' And she did.

Since the beginning of my time as a teacher I had been critical of the effect of the Oxford and Cambridge selection procedure on the teaching in schools. I believed that many would-be scientists concentrated so strongly on science and mathematics for the scholarship examinations that whole areas of human experience were not touched. Manchester Grammar School, for example, provided excellent specialized teaching, and those who could profit from it and gain an Oxbridge scholarship would often, in the atmosphere of those universities, experience other parts of human culture as undergraduates. But we at Bristol got the boys who had been through these specialized courses and had not made it to Cambridge. I did not think much of some of them. I used to ask candidates for admission to our department, 'Can you tell me why the sky is blue?' I felt that it was those who had been drained by working too much for a specialized examination who answered, 'Oh, Sir, that wasn't in the syllabus'. Those who had a shot at it and showed some interest were the ones I wanted to take. When I went to Cambridge in 1954, this experience was very much in my mind, and in Chapters 12 and 14 I shall describe what I tried to do about it.

When I first came to Bristol in 1933 my colleague Herbert Skinner had described our laboratory as a first-rate physics department in a third-rate university. Whether this was fair is difficult to judge, but it is certainly true that the size and resources of the physics department were disproportionately large compared with the rest of the university, and yet we in physics were in no way encouraged to take a main part in its activities. After the war, on the other hand, Morris inspired us all with the ambition to become outstanding among the modern universities. We were rapidly expanding; in contrast to pre-war days, we had students from all over the country. Some from the north said they chose Bristol to get as far away from home as possible, but others had heard that Bristol University was a good place to work in. We certainly had talented students, though perhaps fewer than Cambridge from independent schools and homes with an established university tradition. Certainly, when we later moved to Cambridge, Ruth and I found Cambridge students more at ease when we entertained them at home. However, our students were very satisfying to teach, and several now head physics departments in various parts of the country.

The organization of the university, with its faculty boards, heads of departments, deans of science, arts, engineering and medicine, meant that the responsible people had to keep in touch with all their colleagues and had to know what was wanted; but in the end, with the Vice-Chancellor, they made the decisions. It was up to them to make the university a happy and united place. I was for a time Dean of the faculty of science and thus a member of the Committee of Deans, which was our true 'inner cabinet'. It was Philip Morris who persuaded me that I could and should do this. Later, in Cambridge, he pushed me into some Government jobs, which, perhaps, I did not really have time for, as I shall describe in the next Chapters. He was the man who gave me a taste for committee work and policy-making, and persuaded me that I was better at it than perhaps I turned out to be a decade later. But whether this was so or not, I remember him as one of the most remarkable men I have ever met, and I was deeply grateful for his friendship.

At the end of the war one of our new appointments was Donald McGill, whose job was to look after the many mature students who came back from the war. McGill was not a research man, but a teacher, and when this group of students had been through the university, he resigned and went back to school work. He told me that a place so orientated towards research would not be for him, and that he would feel a second class citizen among us. Later he played a major part in the Nuffield project on school science, where I came to appreciate his merits still further. I was sorry to lose him, and sorry that the university—in his view—did not think as highly of first class teaching as of good research.

The balance between teaching and research has always been a problem for me. I thought and still think that a university should be flexible enough to lighten the teaching load of brilliant men at the height of their creative powers, but at the same time to give credit and promotion to its best teachers. I do not believe that it is necessary for a successful teacher to be engaged in research, except when teaching students in their final year. In Cambridge in the old days the college posts, particularly the tutors who were responsible for advising students on matters outside their academic disciplines, provided avenues for advancement. I do not think that these posts still carry as much prestige as they used to.

My life-long interest in scientific publishing dates from my return to Bristol after the war. Since the pre-war period the number of scientists and the number of publications had very greatly increased. With the near destruction of German science by Hitler and by the war the former pre-eminence of the German journals, the *Zeitschrift für Physik* in particular, had almost disappeared. English had become the language of science and the American journals, particularly the *Physical Review*, published by their Institute of Physics, had cashed in on it. I felt that we should have our share. In pre-war days our most prestigious journal had been the *Proceedings of the Royal Society*, but this, covering all physics and mathematics in one journal, seemed to many of us too wide to meet the needs of the greatly expanded scientific community. In addition we had the *Proceedings of the Physical Society of London* and the *Philosophical Magazine*. The former had a reputation for the publication of sound papers in classical physics, but had not been used in the fashionable fields of quantum theory and the nucleus. I tried to get it changed and to challenge the *Physical Review*; but the people running it were not sympathetic. Then someone suggested that I should take on the *Philosophical Magazine*, founded in 1798 and, at one time, our most important physics journal in which Bohr had published in 1913, but which had now fallen on hard times. I did this, and though I certainly could not make it a rival of the *Physical Review*, my relationship with the publishers Taylor & Francis Ltd. has continued until now, and will be described later in Chapter 19.

During this period I served as president of the Physical Society. My chief task there was to promote the amalgamation between the Society and the Institute of Physics. When this was done, the joint body could and did arrange a proper medium of publication—different branches of their *Journal of Physics* for each branch of the subject, solids, nuclei, atoms and the rest.

A letter at this time reminds me of a problem that worried me in Bristol, as it did later in Cambridge, namely the relation between mathematics and physics in the organization of teaching. I wrote to my parents

October 9, 1948

... I went for a walk with our reader in pure maths—destined for a chair when our professor retires at the end of this year. The important question is, what to do about applied mathematics. The logical thing to do would be to abolish the subject and leave it to the physicists and engineers to teach. But of course the mathematicians are always opposed to that—and perhaps they have some reason. In our university, unlike most, the engineers now teach their own maths; and it's generally felt, for their mathematicians to live in an exclusively engineering atmosphere is a bit narrowing. It is certainly going to be a very complicated question.

During the war contacts with scientists in Europe had of course broken down, and to help in re-establishing them involved me in much travelling. I remember being invited to Holland in 1946 by the students, and experiencing crowded trains, bridges down and roundabout routes by car over temporary bridges. I went to Leiden to see H. A. Kramers—one of the fathers of the quantum theory and a man of the highest stature. 'Now we feel liberated', he said, when he saw me. He took me for a walk along a canal near his house. 'The trees have all been cut down by the Germans, as they restricted their field of fire', he said bitterly.

I also went to Germany in 1946, on behalf of the Control Commission; their scientific branch was at Göttingen. I believe there was some idea that we had to prevent their scientists from getting involved in military research, but this was of course nonsense. Our real object was to help them to get to work again. What I remember most vividly was that, at least among the physicists, there was not the slightest sign of resentment at the saturation bombing or at anything else. For part of the time I was with R.V. Jones. 'Here come Mott and Jones', was the greeting more than once, referring to my book on metals with Harry Jones. This did not please Jones very much; he launched into a description of the electronic techniques used to raise a fire storm in a neighbouring town, with our German colleagues hanging on every word. There was a party at the house of the officer in charge of the scientific section of the Control Commission; Heisenberg was there and other physicists, and also a pilot who had been on the raid in February 1945 that destroyed Dresden. The conversation was all on the technology of destruction. The only resentment I encountered was when I called at the house of Richard Becker, a physicist whom I had known before the war and who, when war broke out, had sent me a message of friendship through a neutral country. The last bottle of wine came up from his cellar; we talked intimately and he told me that he now realized that the stories of Dachau and the rest were not just allied propaganda but only too true. Just then his wife came in and looked at me in a less friendly way and at last came out with, 'Professor Mott, unheard of things are happening in Göttingen (*unerhörte Sachen*). 'What things?' I asked. 'The Americans', she said, 'The Americans! They have commandeered the house of a *Professor*'. Poor Becker, with the taste of the concentration camps on his lips, hustled me out, and when I came back next day I heard nothing further of this.

I went on to Hamburg and saw the results of the first fire storm; I got in touch with one of their scientists (Professor Bagge) and invited him to lunch with me at the military mess where I was staying. The (German) young woman at the door said, 'This is for allied personnel only, you can't bring in a German'. I said, 'We are on official business, see if you can get permission'—and she came back and said, 'you can bring him in, as long as you don't talk any German'. I thought of myself being ordered to talk no English in the Grand Spa Hotel in Clifton if Germany had won the war, and since moreover his English was less adequate than my German, I said, 'This is too bad, let's see what we can get in the town on the black market'. But he, perhaps wisely, had no pride in the matter, came in and lunched with me and talked English, of a kind. From Hamburg I went on to Brunswick and was impressed to find that the first building to have been rebuilt among the ruins was the opera house.

I was also able to visit Paris as early as 1947 and in 1950 and later gave some lectures on solid state physics. French solid state physics had not been able to develop much during the German occupation and my lectures were well received. They were in the laboratory for *Chimie Physique*, in the rue Pierre Curie; this mixture of physics and chemistry, started by Jean Perrin, seemed to me very appropriate for solid state physics. Particularly the work in Paris on soft X-ray spectroscopy under Professor Cauchois (compare with Chapter 9) interested me; I made some contributions to it and kept in contact for many years.

I also gave some talks on subjects other than physics, as the following report from the New York Herald Tribune Paris edition shows (31.1.50) . . .

Science and Truth

The distinguished English physicist, N. F. Mott, F.R.S., of the University of Bristol, paid a visit to Paris where, in addition to giving eight lectures at the Sorbonne on the subject 'The electronic structure of matter', he found time to speak before a small philosophical group on 'Modern physics and the nature of scientific truth'.

In the opinion of Professor Mott, although he considers science to be the 'glory of our epoch', it remains nevertheless an 'instrument for changing the world' rather than one with which 'to seek truth'. He cited as example of its limitations the questions asked by the chorus in the 1st act of T. S. Eliot's *'Family Reunion'*.

'What is happening outside of the circle?
And what is the meaning of happening?
What ambush lies beyond the heather
And behind the Standing Stones?
Beyond the Heaviside Layer
And behind the smiling moon?
And what is being done to us?
And what are we, and what are we doing?
To each and all of these questions
There is no conceivable answer.'
to which he agreed.

But for Mott there exists a more urgent question to which we must soon find an answer, namely, how much suffering may we impose on humanity in order to obtain the political or economic order in which we believe?

In conclusion, he quoted his English colleague, J. Bronowski, who has declared that when science can offer nothing to replace ethics, the result is a series of such clap-traps as: 'enlightened self-interest, the greatest happiness of the greatest number, survival of the fittest, etc.'

Mott was introduced by Gaston Bachelard, professor of History and Philosophy of Science at the Sorbonne, who is the author of some of the most original contributions to French philosophical literature that have appeared since the war.

Soon after the war, in 1947, I returned to North America, first visiting the Atomic Energy Establishment at Chalk River in Canada. Then I went to Kodak's at Rochester, New York, to talk about my work with Gurney on latent image theory, and to the General Electric Laboratory at Schenectady, which under Dr Hollomon was becoming an important centre for solid state physics. I had crossed the Atlantic in a converted bomber but returned on the Aquitania, with a cabin to myself, which was arranged by the Atomic Energy Authority and was an almost unheard of luxury at that time. To mention it to other passengers was to court unpopularity.

With the war over, most physicists were deeply conscious of the problems posed by nuclear energy; we felt it our duty to inform the public about the benefits that we believed could be obtained from nuclear power, but even more to warn of the dangers from the nuclear weapons. My feelings about the news of the bomb are expressed in the following letter to my mother.

Bristol
August 1945

Dearest M,

Well, the cat is now out of the bag about the Pandora Box that we have opened—the unexpected child that was born of the researches of Rutherford and Bohr. The only hope for the world is that the statesmen of the major countries—knowing that a war in, say, 1965 would mean the instant annihilation of their chief cities—will get together and see that it doesn't happen. I know that this is the argument with which Bohr comforts himself.

I have not been in on this except in the very early days of 1940 when the struggle was to get the thing taken seriously. The possibility of the bomb was then common knowledge in physics labs. Recently I have known that it was nearly ready.

I expect you will understand a lot in the light of this, the cancelled Russian visit for instance, in which this probably played a part.

Well, the United States now has the opportunity to enforce world peace—in theory!

In 1946 we formed the Atomic Scientists' Association to put before the public the true facts and to investigate proposals for the control of nuclear energy. I was its first president; Blackett, Cockcroft and Lord Cherwell were vice-presidents. Perhaps I was chosen because I had not been in the Manhattan project. We organized the 'atomic train'—a railway carriage with an exhibition of what could be shown about the physics of nuclear energy. We started the *Atomic Scientists' News* in 1946, which was published from 1951 by Taylor & Francis. I spoke frequently on the subject, and in many places, and also wrote several articles in the *Atomic Scientists' News* and elsewhere. Margaret Gowing, the historian of the British Atomic energy project, writes in Vol. 1 of her *Independence and Deterrence* (ref. 9, p.184) that the decision that the UK should make the bomb was accepted almost without question, and that in the *Atomic Scientists' News* the only vociferous opponents of the bomb were Dr. Kathleen Lonsdale, a lifelong pacifist (she was a Quaker) and Professor N.F. Mott.

Below are reproduced two articles of mine from this journal and from *World Affairs*. Though I did not anticipate the hydrogen bomb, I was not otherwise far wrong about the future.

International Control of Atomic Energy
The Choice Before This Country
By N. F. Mott

In the autumn of 1947 it is idle to pretend that the chances of obtaining international control of atomic energy in the near future are good. Nevertheless it remains as important as ever to work for international control, even though success will take much longer to achieve than appeared possible two years ago. It is therefore necessary for us in the Atomic Scientists' Association to consider what steps can be taken which will make the attainment of some of our aims less unlikely. Unfortunately we must also define our views on the policy that should be followed by this country if, as appears only too probable, no system of control is set up in the next few years. Consideration of these two issues is the purpose of this article.

Let us at the outset be clear that the aim of our policy is to prevent war. It is not simply to abolish atomic bombs while retaining other weapons; we do not believe that this is possible, because atomic weapons are so effective that they would almost certainly be used in a conflict between great powers, if their use would give a clear advantage to either side. Atomic weapons are relevant to the problem of the abolition of war for two reasons; the first is that their widespread use in a major conflict is likely to be so destructive to both sides taking part that it has become even more important to avoid war than it was before; the second is the common interest that all countries have for avoiding atomic warfare, which may well act as a powerful incentive for doing something to avoid it.

We, in this vulnerable and densely populated country, have as great an interest in peace as any other people; in calling for a crusade for international control among the people of Britain the Association would therefore be preaching to the converted. It is not a crusade that is called for at present, so much as hard thought about the steps that ought to be taken, together with much educational work to bring home to the country the true facts. But before giving an analysis of the position and suggesting certain steps, it will be well to dispose of one course of action which any officer of the Association and indeed any scientist must have heard proposed scores of times. This is that scientists hold the key to world peace in their hands, and that if they refused to work on atomic energy or on atomic weapons the governments would not be able to make this kind of war. This is undoubtedly true; but in fact the scientists of England and America are no more likely to behave in this way in times of national danger than are the men of the armed forces, or the makers of munitions; as for the scientists in Russia, we have almost no contact with them, and cannot tell what they would do, but we can guess that their behaviour would be the same. Refusal to work on atomic or other weapons is a course which some men take as a result of their views on right and wrong; it is a course which, if adopted by a large number of men in this country, might have some political influence, but which does not seem certain to make war less probable.

International Control: The Present Position
The Lilienthal report on the control of atomic energy was rightly welcomed in this country as a great and inspiring document, and its acceptance by the American State Department as the basis of their proposals to the Atomic Energy Commision of the United Nations seemed to many of us a real step forward in the history of statesmanship. It will be well to recall the outstanding points in the report. These are:

(*a*) that an international 'Atomic Development Authority' should be set up which should own all uranium mines and all power plants capable of making fissile materials (the raw materials of the bomb) in dangerous amounts.

(*b*) that the authority should carry out research and development in atomic energy, in order that it should offer an attractive career to first class scientists; without this it would be impossible to exercise effective control.

(*c*) that the Authority should have powers of inspection, to ensure that no government was making fissile material secretly.

In their plans or dreams for the distant future, most members of the Association still feel strongly that a constructive scheme of this kind is to be aimed at, and that its setting up would represent an achievement of the first order for world peace. In considering short term or immediate steps, also, it is necessary to have an ultimate aim in view, and this the Lilienthal scheme provides. But unfortunately agreement on this scheme appears at the moment most unlikely. The reason for this is, of course, the friction between the United States and Great Britain on the one hand and the

U.S.S.R. on the other, which is the outstanding fact in the world situation today. Therefore any immediate proposals for the control of atomic energy must be made with a view to lessening this friction, especially because it is only in an eventual conflict between these powers that there is any serious risk of the use of atomic bombs in the next few decades.

In any plan for control, therefore, which involves some renunciation of national sovereignty such as is implied by ownership by the Authority of mines and factories, a situation must be envisaged in which either the United States or else the U.S.S.R. is asked to obey a decision of the Authority on some vital matter. For instance, according to the proposals of the Lilienthal report, the A.D.A. would have to decide on the allocation of raw materials (uranium) and on the distribution of piles between the countries concerned. It is clear that the U.S.S.R., whether justifiably or not, regards the world as being split fairly sharply into countries favourable to them, and countries grouped round the United States. It is therefore hopeless to expect their government to accept as binding any decision of an international body on such vital matters. The United States would take the same view if the number of states voting consistently with the U.S.S.R. exceeded the number voting for them. This point of view has been expressed by Mr. Gromyko himself, who said: 'The Soviet Union is aware that there will be a majority in the control organ which may take one-sided decisions, a majority on whose benevolent attitude towards the Soviet Union the Soviet people cannot count. Therefore the Soviet Union, and probably not only the Soviet Union, cannot allow that the fate of its national economy be handed over to this organ.'

With the world divided into two blocs, then, an international body is likely to reach decisions only if the men and women who form part of it owe loyalty to it, rather than to their sovereign states. We may hope that such loyalties may arise in the future. At present they do not exist. It follows, therefore, that at the moment it is futile to advocate the setting up of one authority which will control the activities both of the U.S.S.R. and of the U.S.A., either in atomic energy or anything else. No body can be envisaged whose impartiality and justice the governments of the two countries would trust. Therefore the effort to be spent in each country on atomic energy and the location and type of plant must be left to the national governments concerned, and we must regretfully admit that any attempt to have it otherwise will not succeed in the present state of the world.

If this analysis of the situation is correct, it appears that the divergence between the Soviet viewpoint and the American proposals put forward by Baruch will not easily be bridged. The American statesmen and scientists have put great emphasis on complete control and this is more than the Russians can accept. Nevertheless, it is still true that the only realistic policy for this country to try to reconcile the viewpoints of America and Russia; the following very tentative suggestions are made to suggest how this might be begun.

First Steps for Establishment of an International Authority
If the concept of control is left out of the Lilienthal scheme, there remain two things; the idea of an international body of scientists charged with the development of atomic energy, and the proposals for inspection. It seems possible that with this residue something of value can be built up. Let us consider the question of inspection first. On this the attitude of the Soviet statesmen has been by no means uniformly hostile, although we can be sure that they would not admit inspectors without very great inducements. It cannot be denied that the admission of inspectors would decrease the military strength of the U.S.S.R. Under present conditions little is known of their new centres of heavy industry beyond the Urals, or of the state of their atomic research. Inspectors who were nationals of a foreign power would obtain such knowledge, which would be of undoubted value to the chiefs of staff of the foreign power concerned.

From the American point of view the outstanding fact in the situation is that the discovery of the bomb will abolish in a few years' time the security of their homeland from sudden attack. The bomb for them, in the 1950's, is like the development of aviation was for us in the 1930's. This has been realised by American public opinion. Therefore there is widespread demand for security through control and inspection, and willingness to surrender perhaps even some degree of national sovereignty to obtain it.

It would probably not now be agreed by most Americans that by inspection alone any worthwhile security could be obtained. Nevertheless it seems to the writer that along lines such as these resides the best hope, even if at the moment it appears only a small hope, of bringing about a reconciliation between the points of view of America and Russia. Let us suppose that there were established in various countries, including these two, laboratories for research on atomic energy in each of which outstanding scientists from the western and eastern powers would be invited to work. Let us suppose also that the governments concerned would undertake to give representatives of these laboratories access to their mines and plants sufficient to ensure that no unreasonable stockpiles of fissile material were being made. In this way America, and for that matter Russia, could have access to information that would show them that no preparations for sudden attack were being made. The resulting lessening of suspicion could be very great. In fact the degree of security obtained would not be much less than under a full-fledged Lilienthal plan even armed with the abolition of the veto, for in no case is there anything to prevent a great power, dissatisfied with the decision of the Security Council, from disregarding its orders and seizing such plants as exist on its territory and starting to make bombs. It must always be remembered that, with the balance of power as it is today, no sanctions can be applied against America or Russia without involving the world in a major war.

It will be argued also that the inspectors or scientists in Russia would be hedged around by so many restrictions that they would be unable to obtain the information that the rest of the world would desire. It is of course true that no scheme will succeed if the governments concerned are determined that it shall not. We have to decide whether, in the present state of the world, it is worth making any proposals at all.

Any scheme for the international control of atomic energy, set down coldly in black and white, looks unattainable in the present world of suspicion and power politics. Nevertheless, in the writer's view, it is necessary to make proposals for such schemes, because in atomic energy the world is confronted by a new situation. These proposals are put forward as basis for discussion in the belief that they form the minimum that would do any good, and that they should be put before American public opinion. It is only America that can make an offer along these lines. What the Russians would demand in exchange it would be for them to say. They might ask for the cessation of stockpiling of fissile materials, the outlawing of bombs, for technical information or even for a dollar loan. The issue could only be settled by hard bargaining between the two countries.

Policy for this Country if no Agreement is Reached
If no agreement is reached between America and Russia, we may envisage that Russia will continue to make every effort to manufacture atomic bombs and to obtain favourably disposed governments in the states on her borders. In America the development of atomic weapons will continue, with its accompaniment—the militarization of science and the imposition of secrecy on a greater and greater proportion of their research. In addition, the importance of overseas island bases for the United States will increase. The time

is still far distant when Russia and the United States can attack each other from home bases with automatic weapons of the V2 type, and for attacks with piloted aircraft, bases near the objective will be important to avoid prohibitive losses. These points are well brought out in two recent semi-official articles on United States Navy and War Department thinking on the atomic bomb. Both for offensive and defensive purposes, the value to both sides of allies and bases distant from the homeland will be very great, as a sort of screen or cushion that will absorb the first shock of an atomic war.

In this situation, what should be the policy of this country? It is not the role of the Atomic Scientists' Association to advise on whether we should ally ourselves with America, whether we should attempt to form a western European bloc or whether we should attempt to follow an independent course in friendship with Russia. Some facts may, however, be pointed out. In alliance with America, nothing that that country could do would save us *if* war should break out against a power capable of occupying the channel ports and equipped with atomic bombs. Fifty of these missiles, launched with V2 weapons in the present state of development, could kill a quarter of the population of London and make the city uninhabitable. An alliance would, from the point of view of this country, have value only in the hope that its deterrent effect would be great enough to prevent war—to establish a pax Americana. As recently pointed out in an article in the *Bulletin of the Atomic Scientists*, the deterrent effect will only be great if the alliance remains strongly armed with atomic weapons and with a highly developed passive defence, with all that that implies in interference with the life and liberties of the countries concerned.

Whether or not we favour the American alliance, there are certain decisions in the more limited field of atomic energy which will have to be taken by this country. At the present time large sums of money are being spent by the Ministry of Supply in the development of atomic energy at Harwell and the North of England. This can be justified on three grounds:

(*a*) That invention of chain-reacting piles is a revolutionary scientific discovery of immense importance, and it is essential to develop them in this country, in order that we may keep our scientific pre-eminence with all that that implies for our industrial developments. The provisions of strong sources of neutron radiation and of tracer elements would alone justify some development.

(*b*) That atomic energy offers us hope of a substantial addition to our fuel supplies, at a date in the future which is difficult to predict and on which various authorities give conflicting estimates.

(*c*) That in a few years' time we shall be able to make atomic bombs.

In the development of any project for atomic energy, the initial steps that must be taken will not depend on which of these ultimate aims is in view; certain technical problems must be solved and a large pile built. It is safe to say that the time has not yet come when any definite decision must be taken. But it will come in this country soon enough. It will come in particular when the first pile begins to produce fissile material (plutonium) in significant quantities. What is then to be done with this plutonium? Two courses are open to us. We can store it up in underground places, ready for use in bombs if need arose, and concentrate all our energies in building more piles to produce more and more of the material. Or alternatively we can devote all our energies to the production of power. In this case the plutonium will be fed back into a different design of pile where in the end it will be used up, so we shall never accumulate a large store. These two courses involve somewhat different engineering projects; there is a real choice before us. The decision is one that the government will have to make after considering the views of its political, military and scientific advisors. The military, whose duty it is to state the requirements for defence in the event of a war, are almost certain to urge the manufacture of bombs. If the majority of scientists decide that they are opposed to this policy, they must oppose it actively and secure national support.

Some general arguments that can be put forward for accumulating bombs are as follows: If we had some we could threaten retaliation against any power that might use them against us; they would strengthen the hand of the Foreign Secretary in international affairs; they are the weapon of a great power, and if we do not have them we renounce our position as a great power. These are all arguments connected with prestige.

The following reasons can be put forward against the policy of making bombs: The only countries against whom we could either use or threaten the use of atomic bombs in the foreseeable future are the U.S.S.R. or possibly a resurgent Germany which could only come into being as a threat to us under Russian influence. Now it is entirely unrealistic to suppose that we can even consider atomic warfare with Russia with any chance of success without full American support. Even on the doubtful assumption that we would have a greater supply of bombs, we are infinitely more vulnerable. The Russians could do us more harm with ten bombs than we could do to them with a hundred. But if we have an alliance with America, it is obviously preferable for military reasons that bomb production should be concentrated in their less vulnerable continent just as it was in the war which has just ended. This is almost certain to be the view of the Americans themselves.

But suppose that we do not have an alliance with America, either because we cannot have one on acceptable terms, or because an alternative policy proves preferable. We have to ask whether atomic bombs would be of any value to us then. We have again to face the fact that any power occupying the continent and possessing atomic bombs is in a position to destroy us. In that case, what should our defence policy be? Clearly it should be designed so that no power should have much to gain by attacking us, but that any power attempting to occupy our country (perhaps to use as a base or deny the use of our territory to an enemy) will lose heavily in the attempt. If we had atomic bombs, we could say that if attacked we could retaliate on such cities of the enemy as were accessible, though probably we could not do much harm to centres behind the Urals. Also atomic bombs could be a very effective weapon against any attempt at seaborne invasion involving the use of harbours. These are sound military reasons for possessing bombs. The reasons against possessing them are less tangible, but none the less real. One of them is that, at some not too distant date, when the technical information we had from America before 1945 is no longer of importance, our project for atomic energy could cease to be secret. It could be carried out in collaboration with the countries of Western Europe and indeed with any country that was willing to work in the open. We could let it be known that we were making no stockpile of bombs with which to attack anyone, and we could give the Russians and everyone else the opportunity to come and see that this was so. We could attempt to make atomic energy a field for collaboration between the nations instead of a source of suspicion.

Citizens of this country may well have to decide which of these policies is the right one; whether the moral lead which the peaceful and open development of atomic energy would give outweighs the confession of military weakness which it entails. Such matters cannot be decided on scientific grounds alone. It is the duty of the A.S.A., representing the informed opinion of scientists in this country, to state the facts on which a decision must be taken.

(*Atomic Scientists News*, 7 Sept. 1947)

The Atomic Bomb and World Affairs
By N. F. Mott

It was widely believed before the war of 1939 that the first days of hostilities between heavily armed Powers would see important cities on both sides reduced to ruin by aerial assault. The war itself proved that the destruction of cities was a harder task than had been imagined. The German air force, designed to support the army, had some successes against lightly defended towns, especially before adequate fire-fighting services were built up; but it was never large enough to achieve destruction on a scale that would affect the course of the war. It took four years, during which a very high proportion of the scientific and industrial effort of England and America was allotted to the task, to produce the massive air fleets which could destroy Hamburg in a week or Dresden in a night. The more spectacular attacks, too, both on Germany and Japan, took place when the power of defence of the enemy was beginning to weaken and when his power to hit back, except with V weapons, had almost disappeared; we have never had, in this war, two countries of almost equal power hitting at each other's centres of production to see which side could stick it out the longest.

In spite of the prophets, therefore, and in spite of the influence of air power on naval and military operations and on German industrial production, the European war was won ultimately by soldiers who broke through the enemy's defended positions and occupied his country. Neither bombing nor the threat of further bombing led to a break in civilian morale great enough to compel surrender.

With the coming of the atomic bomb, it has been widely stated that the days of soldiers' wars have gone, and that if another major conflict breaks out, the issue will be decided by the atomic bomb alone, leaving the territories of both combatants far more exhausted and destroyed than in Europe to-day—or in 1918. It is very important to know whether this estimate ought to be compared with the more alarmist forecasts of the effect of air power made before 1939, or whether, between heavily defended powers, the bomb will be a terrible weapon indeed, but not one entirely decisive or precluding the existence of an industrial organization enabling armies to fight. Some estimate must therefore be made of the probable development of the bomb and of its military potentialities.

The bomb dropped on Hiroshima must have been a 'mark I' model and can doubtless be improved; however, the amount of energy which can be released from uranium or from similar materials by fission was well known even before the war; and if the published accounts of the effectiveness of the bomb are accepted, it is clear that it is already highly efficient, and it is doubtful whether further research will increase the effectiveness of the explosive by more than two or three times. In this respect the development of atomic explosive is likely to be similar to that of the bomb fillings of this and of the last war, amatol, T.N.T. and the rest; here the most modern materials are only about twice as effective as those of the last war. It will, of course, be possible to make larger bombs, but in view of their extreme destructiveness, it would probably be more efficient to use a large number of small ones instead of one very large one. Research is likely to be concentrated not so much on improving the bomb as on cheapening the method of manufacture. The separation of uranium 235, or the transmutation of uranium into plutonium, will remain a fairly expensive process; it is just possible, however, that cheaper materials will be discovered.

The effects of the bomb have been stated to be equivalent to those of 10,000 tons of T.N.T.; this weight of explosive could actually be dropped by Bomber Command in three or four raids, and would be enough to destroy a large town. With the atomic bomb, however, the degree of destruction in the centre of the area of devastation would be much greater than anything that could be achieved with T.N.T.; air-raid shelters would be destroyed and the loss of life would be very heavy. As against this, destruction is unlikely to be spread over such a large area as after attack by a large number of T.N.T. filled bombs.

The effect of the bomb on strategy will depend entirely on the productive capacity necessary to make it, and thus on the number of bombs with which a country could hope to start a war. It is highly unlikely that the material will ever be produced in sufficient quantities to be carried by the massive air fleets of to-day. One pictures attacks by single light bombers each carrying one bomb. But even a production of, say, a few hundred bombs a year would soon enable a store to be built up which would have a devastating effect on any modern industrial state if they could be safely delivered to their targets. Although improvements in anti-aircraft defence may make it more difficult for the attacking bomber to get through, the small weight of the atomic bomb and consequent speed of the aircraft, and the possibility of delivering the bombs by rocket as well as from aircraft make the hope of any complete defence remote. The strongest country will be that in which the centres of population are most dispersed, so that there are few targets worth the expenditure of a million-pound bomb.

In fact, dispersal of industry and population would be the first preoccupation of a government which took seriously the defence of its country against atomic warfare; the number of bombs which any enemy will be able to drop is likely to be severely limited, and if industry was scattered over a sufficiently large number of centres, something might be left at the end of the attacks. Ports and har-bours cannot be dispersed, however, and they would be particularly vulnerable, as would be also the large capital cities if allowed to remain in their present form. Germany, with her major cities largely destroyed, may well attempt to rebuild her industry in much smaller centres with atomic warfare in view.

Unless dispersion can be achieved to an unheard-of extent, atomic warfare is likely to bring suffering and destruction to a much greater extent and much more rapidly than in the war which has just come to an end. There is, however, one hopeful feature of the picture; it will no longer be possible for an aggressive Power to hope to achieve a cheap victory over its neighbours. In 1939 the Germans must have known that the opposing air forces were not strong enough to inflict substantial damage on their country, and they believed, not without some justification, that their army could invade and conquer any European country before its air strength could become dangerous. But in a future war, if two or three of the enemy's fast light bombers can get through to your capital, your capital is destroyed; you may bomb his airfields, send up your night fighters, concentrate your A.A. defences, but you will not make your capital safe; a few bombers will get through. In fact, if the enemy had bases within a few hundred miles of your capital, the bomb could even now be delivered by rocket and not by aircraft at all; in the future the ranges of rockets will increase.

Some years will pass during which the bomb will remain presumably in the hands of the United States Government only; but the general principles of the bomb are common knowledge, and doubtless in a few years' time it will be in production in other countries. It is highly desirable in the interests of science that all the physical principles involved shall be divulged to the world; a whole new chemistry, 'neutron chemistry', must now exist in relation to these atomic processes, and the information about it is hidden away in secret reports. But whether or not research on atomic power is carried on in secrecy or in the world of international science does not seem to make much difference to the possibilities of atomic war; what will matter for power politics will be the capacity for manufacture of the atomic explosive and the store of bombs available.

International control of the manufacture of atomic explosives has also been suggested. While this is obviously highly desirable, it is only fair to point out the difficulties. We are entering an age of atomic power; these same materials, uranium 235 and plutonium and any others that may be found, will have industrial applications that we cannot at present foresee. International control of the explosive will mean international control of a gradually increasing part of the economic life of a nation. In the end it would lead to the abolition of a large part of national sovereignty.

We can only hope that, with this sombre bomb in the background, war between nations will come to be regarded like war between Canada and the United States, as something unheard of, against which no preparation need be made. This, or a world state, seem the only alternatives before civilization.

(The Quarterly of World Affairs, London Institute of World Affairs, Vol. II, Oct. 1945)

I also reproduce the minutes of a meeting in September 1951, when we all crossed swords with Lord Cherwell about our role in politics.

The discussion at the meeting turned on the policy to be adopted by the Association towards matters having political implications. It was opened by Professor Lord Cherwell, who urged very strongly that the Association should refrain altogether from expressing views on political matters. Scientists were not united in their political convictions. He felt that the most useful role for the Association lay in giving expert guidance and assistance to the public and to the Government in technical matters such as, for example, Civil Defence arrangements. Expressing views on whether atomic bombs ought or ought not to be used in warfare was not our job, and could do harm. Bodies like the Royal Society did not indulge in politics: why should we?

Professor Mott, voicing the general attitude adopted by Council during the year, agreed with Lord Cherwell that it was impossible for scientists as a body to agree on political matters, and that the Association ought not to make political pronouncements. But he did consider it important for scientists to have a forum on which they could discuss amongst themselves and with the public the social and political implications of their work. Professor Rotblat recalled that in its early days the Association was preoccupied with the question of International Control—a problem which certainly embraced both science and politics. Other speakers also urged the importance of discussing the social implications of atomic energy, and therefore of discussing politics to a limited extent. According to Mrs. Simpson, a member of the editorial staff of the American *Bulletin of the Atomic Scientists*, whom we were glad to welcome as a guest to the meeting, the practice of the *Bulletin* is to devote about two-thirds of its space to social and political matters. It is very careful, however, to include in the same issue articles giving both sides of any controversy. In this way it hopes to receive criticism equally from the right and from the left.

Professor Mott welcomed this definition of impartiality, and suggested that the *News* should try to emulate it. However, Lord Cherwell did not agree. He deprecated the idea of conducting a sort of political forum in the *News*. Scientists wishing to express political views should join frankly political organisations, and allow the A.S.A. to remain as a learned society that could give the public reliable and unbiased statements about atomic energy. It would be very difficult to maintain a just balance between left-wing and right-wing views in the *News*: experience showed that while left-wing writers were prolific of articles, the right-wing did not have the time to contribute regularly to a journal with such a small circulation.

Commenting upon Lord Cherwell's statements, Dr. Kurti remarked that political aspects of scientific questions are touched upon in the editorials of *Nature*—could we not do likewise?—and Professor Mott pointed out that the small circulation of the *News* did not truly indicate the extent of its influence. One of his articles had subsequently found a place in the *U.S. Army Infantry Journal*.

Lord Cherwell's point of view was supported by other speakers, who suggested that the *News* could be run effectively as a journal of popular education, with articles on the uses of isotopes in medicine and industry, etc. The Editor, while agreeing that such articles were valuable, pointed to the difficulties in securing a flow of articles of this type, and asked whether the enthusiasm of those doing voluntary work for the Association could be maintained on such limited objectives. In dealing with controversial matters his aim had been to maintain a proper balance between all points of view, but—as Lord Cherwell had already pointed out—it was not easy to find the people to write the article.

The discussion did not end with any agreed conclusions. Summing up, the President indicated that the new Council would of course take note of the views expressed. In future issues of the *News*, he thought that care should be taken for articles not to digress from science and its relations with society into politics as such, while in all controversial issues the attempt should be made to represent the conflicting points of view in one and the same issue.

Before the Atomic Scientists' Association was formed, I was a member of the Atomic Scientists' Committee of the Association of Scientific Workers, together with Blackett, Bernal and Burhop. Later I think that my political views must have moved a good deal to the right, as is shown by an extract from a letter to my mother, dated November 1948, about Desmond Bernal, the leading and very influential scientist of the far left.

> We had Bernal down to address the Association of Scientific Workers (A.Sc.W.) of which he is chairman. I took the chair because I don't think anyone ought to get away with his impudent assumption of peace as a communist speciality. My remarks from the chair were listened to in stony silence but provoked Bernal into a passionate defence of Soviet policy which was just what I wanted. Loud applause for B. But all the same—with some of my young physicists there, it *must* be made clear that Bernal's conclusions (organise a will to peace among the peoples of the west) only follows from his premises (the sole danger to peace comes from the U.S.).
> I think I'll leave the A.Sc.W. I've served 2 years. It's one of these trade unions that come under communist domination. Great pity because it could be a useful link between scientists and the T.U.C. It sends a representative there, who always votes with the left minority...

I also became involved in the international aspects of physics in other ways. At the suggestion of H. A. Kramers, I succeeded him as the president of the International Union of Pure and Applied Physics (IUPAP) in 1950, an office I held for six years. This organization, with the grand-sounding name, was affiliated to UNESCO and undertook tasks which were perhaps of minor importance. It sponsored international conferences, it had a committee on symbols and units, it advised on physics publications and every three years held a somewhat expensive business meeting. I remember, when I was first appointed, going to Paris to see the secretary (P. Fleury) and asking him what the union actually did; he pointed to his shelves laden with paper and said, '*Voilà*'. At one business meeting the main issue was whether the Republic of China (Taiwan) and the People's Republic (the Mainland) could both belong—an issue which certainly taxed my chairmanship. The further I got from the centres of physics in Western Europe and America, however, the larger IUPAP appeared. In 1953 I went, as chairman, to an International Conference on Theoretical Physics in Japan, which lasted with side trips about five weeks. It was the first major physics conference since the war in Japan, and had enormous publicity in their press; my colleagues and I were invited to speak and broadcast all over the place. At the many banquets (being president) I was always placed between the two oldest and most distinguished of our hosts, who—alas—could rarely speak anything but Japanese. At last I rebelled—could I not sit next to some young Japanese, man or woman, who had studied abroad? Our charming hosts did everything we suggested, and I learned a lot about Japan. Another honour, not entirely welcome, came my way in Sendai. Japanese hotels were excellent and, as usual, each of us, numbering perhaps six or seven, were shown each to our separate room. Each of us, that is, except the two senior members, myself and my colleague from Bristol, Charles Frank, who was travelling with me. Recently the Crown Prince of Japan had stayed in that hotel, and it was accounted an exceptional honour to sleep in the room that he had used. Thinking that the honour should be diffused as widely as could be, they put Charles in with me. Charles is my very good friend—but, with a speech to make nearly every evening and much else, I needed my sleep. Charles' enthusiasm for physics was such that he needed much less and talked till late. I would have liked a room to myself.

Physicists can be like that and do become oblivious to their surroundings. I remember that the first time we went out of Tokyo, to see the temple at Nikko, we passed through a dream landscape, a veritable Japanese print, farmers in straw hats cultivating the rice fields, and the mountains in the distance. But in our carriage I was the only one looking out of the window.

The Japanese physicists, who could all read English, found to their dismay that they could not understand the spoken word. Nevertheless they turned up in their hundreds to hear us. I remember in one place they asked me to give the same lecture that I had given elsewhere. I asked why and they said, 'Because a translation into Japanese has been published in the newspapers'.

We went as far south as Kyoto and north to Hokkaido. The hospitality was lavish, including a Geisha dinner. The Japanese physicists, who were anything but well paid in those days, had never been to such an occasion before; they enjoyed it.

Letters to my wife give something of the flavour of that time . . .

> Imperial Hotel, Tokyo
> Monday 7 Sept.
> *Nothing* can describe our arrival in Tokyo—as soon as we got out of the plane, there were Fujioka and Kishani, the organizers, and about ten (twenty?) reporters, photographers. It was dark—photographed by flashlight over and over again, what did I think of Japan, the purpose of the conference, what did we expect to get

out of it, what was the state of theoretical physics in England? . . .
At length a man from the Japanese travel agency got us away,
shepherded us through the customs, got us into a large car with
Fujioka and Kotani and with more photographs and questions I got
away. Then to the hotel, and after a brief wash away to a Japanese
style meal. Thank God the journey was smooth and we'd had a rest
in Hongkong. The meal was the queerest thing ever, in a room to
ourselves, where we sat along a ledge on the other side of which
were two cooks who fried bits of fish and passed them to us, and
a waitress bowing and smiling and helping us with our chopsticks.
It was very good really but chopsticks are hard work—Charles said
it gave him writer's cramp. Little glasses of hot saki—very good but
one doesn't know how much one can drink safely . . .

Sapporo was the usual thing—two daughters of local notables
and a magnificent bouquet at the station—carnations all wired into
shape—visits to labs, popular lectures by Néel and me, an afternoon where members of the department presented their work,
formal dinner by Mayor of Sapporo, speeches (everyone asks me, do
I propose to publish my collected speeches in Japan?). And finally
today a trip in a bus into the mountains to a lake, lovely blue sky,
wooded hills just going autumnal and the only thing that made it
specifically Japanese was a volcano quietly smoking 2000 ft. above
us. I wish we'd started earlier, cut out the boat trip and climbed
it—but it is not believed that we can walk . . . Sapporo students
who came on the trip too, continually photographing me and asking to be photographed in my company, began to get on my
nerves. By the standards of any other country they would be d.
rude. . . .

In 1953 we did not anticipate the rapid rise of Japan to industrial
pre-eminence. But we did sense optimism and the will to succeed, and
we had the impression of a very orderly society that had renounced
militarism and where no-one would stand in the way of the new
national ambitions. I do not think that much had yet been achieved in
Japanese experimental physics since the war—but I remember the intense will to learn and the very large numbers of young physicists attending our lectures. Yukawa, the Japanese Nobel prizewinner, who
gained the award in 1949 was a theorist.

Travelling in Europe as I did, I experienced the advantages of having some knowledge of French and German; languages seemed to me
grossly neglected in the education of our scientists. When therefore our
professor of French asked me to be president of the Modern Language
Association for a year telling me that my only duty was to give a

presidential address, I accepted. My address (reproduced in an Appendix 1) made my plea for an understanding among scientists of languages and cultures of foreign countries. The entry in *Who's Who* 'chairman of the Modern Languages Association 1954–5' has given me the reputation as a linguist, hardly deserved. It has however introduced me to one interesting experience, which I shall describe in Chapter 12.

CHAPTER TWELVE

Invitation to Cambridge

The first I heard about a possible vacancy in the Cavendish Chair seems to have been when I was in Paris in January 1950, and my mother passed on to me a report which she probably heard from friends in the university world, that Bragg might be leaving Cambridge. A letter to my mother says '...Very interesting rumour about Bragg and the Cavendish ... If so, I wonder whom they will appoint.

I have a curious horror that they might ask me. I don't want to go to Cambridge ...'

However Bragg did not resign until 1953, when he took the position that his father had once held, director of the Royal Institution and Davy Faraday Laboratory in London. Speculation about who should succeed him was a feature of the tea table conversation in my laboratory, as it was, I am sure, in others, even in the United States. The rumour circulating was that the order of preference of the electors was Cockcroft, Bullard, Dee. Sir John Cockcroft, director of Harwell, sat on the fence for a long time; he loved Cambridge, he had kept the house he built there and he wanted to bring the Cavendish back into the Rutherford tradition of nuclear physics. But he had created Harwell, the Atomic Energy Research Establishment, which did a wide range of long-term scientific research as well as work for the atomic energy programme (see Margaret Gowing, *Independence and Deterrence* [9]). He was devoted to it and decided to stay there, and did not leave until he became Master of Churchill College at Cambridge in 1966. Bullard was a distinguished and original geophysicist, Dee the best known of Rutherford's former co-workers. But in the event the electors, when Cockcroft had said no, met again and offered the job to me.

Before I describe my reactions, I must say something about Bragg's reign in the Cavendish and the intense emotion felt by many of my colleagues about the chair and the Rutherford tradition.

Rutherford was the man who discovered the atomic nucleus, who first showed that it could be broken up and in whose laboratory Chadwick discovered the neutron. In 1932, as I have described, he and his team were at the top of the world. But from 1933 onwards his staff began to leave, to replace professors who retired in other universities. Chadwick went to Liverpool, Oliphant to Birmingham, Dee to Glasgow, Feather to Edinburgh and Blackett to London. This was right and proper; these departments were often moribund as regards research and Rutherford's men brought them new life. In retrospect though, perhaps it led to too great a concentration on nuclear physics in the UK, but I cannot remember that anyone said so then. In Cambridge, it was to be expected that their places would soon have been filled by Rutherford's young men. But then, in 1937, quite unexpectedly, after a hernia operation, Rutherford died. A successor had to be found.

It was generally expected that someone like Chadwick would be appointed to continue his tradition. But the electors thought otherwise. Bragg, at that time director of the National Physical Laboratory, had been head of the second most important university school in the country (Manchester); it is possible they thought a change from nuclear physics would be right, but I do not know their motives. But it is certainly true that among many of Rutherford's pupils the appointment was resented. Bragg's achievements were in crystallography, the science of the arrangement of atoms in solids, as far from Rutherford's as they could be.

Since war came in 1939, it was a long time before Bragg could make his mark on the Cavendish. His great success after the war was the appointment of crystallographers like Perutz, Kendrew and Crick who, under his leadership and financed by the Medical Research Council, laid the foundations of molecular biology and the physico-chemical origins of genetics (the double helix). But what to do about nuclear physics must have been a nagging problem to him. At a time when the Americans were spending big money on the subject, and the centre of activity in the UK had passed to Liverpool, Birmingham and Glasgow, the Cavendish was still using some rather antiquated equipment from Rutherford's last days. Bragg was however persuaded—perhaps against his better judgement—to build a linear accelerator for high energy particle physics, which could take the Cavendish back into this field, even though it could hardly be competitive with some of the American installations.

When Cambridge sent me the invitation, I was surprised and, I suppose, flattered. I am a theorist, and this was the Cavendish Chair of Experimental Physics. To step into the shoes of J. J. Thomson and Rutherford was, in my view and that of most of my friends, an overwhelming honour. My mother had recently died—I was sorry she could not know. One does not say 'no' to the Cavendish Chair and I accepted at once. At the same time I was not entirely glad. My Bristol laboratory, highly successful, was something that I had made myself; the key appointments had been mine. In the Cavendish there were various groups, some successful and some less so, and what could I add to what they were doing? No retirements seemed to be ahead; it was doubtful if I could do anything of my own—Bragg had just filled up the last vacancy. And above all—what did the university want to do about nuclear physics? They had just begun on a major nuclear project, and yet they had appointed me, a specialist in solid state. I went to Cambridge to find out.

The Secretary of the Board of Electors to Chairs was the Registrary, who looked after the Council of Senate, the supreme administrative body, but which delegated most matters of importance to the General and Financial Boards. He was in no position to tell the electors what *kind* of professor would suit the department, whether his interests could be afforded, how they would fit in with people already there. So whether they knew or not was a matter of luck. Only much later, after my retirement, was this changed and the responsibility given to the Secretary of the General Board, whose business it was to know those things.

I went to see the Vice-Chancellor, Sir Henry Willink, who had of course been chairman of the Board of Electors. He was charming, but when I asked 'What does the university want to do about nuclear physics?', he sounded as if he had never heard of the subject. He told me to see Harold Taylor—my former student who was now secretary of the General Board—the university's most powerful central committee. Taylor said that the General Board had not discussed it. I went to see James Chadwick, formerly Rutherford's second in command, who had now come back from his chair at Liverpool to be Master of my Cambridge College. He was one of the electors, and explained to me how difficult the choice had been. He said—'I think, perhaps, taking everything into account, we may have done the best thing.'

I realized that there was no Philip Morris in Cambridge. A new professor paddled his own canoe. It was up to me and my colleagues to decide. What was I to do about the linear accelerator?

I felt this machine was a profound mistake. The Cavendish must excel or nothing. What chance had we with this machine, miles behind the Americans? And yet I knew that men I admired thought otherwise. I remember a night in a bitterly cold guest room in Caius College, worrying about it in a sleepless kind of way, and deciding that I must stop it. I am not sure that this was a wholly rational decision—just something that I had to do to prove myself. Certainly it cost me something. But in the event only a few people seemed to mind, and others admired decisive action for its own sake. I think it was the right decision.

I was of course summoned before the General Board, but managed to convince them. They treated me very generously, giving me as 'new professor treatment' a new lecturer in theoretical physics (I appointed John Ziman from Oxford) and the opportunity to invite Hans Bethe for a year from Cornell.

I remember that the day Ruth and I packed the last of our luggage into our car at Bristol, and set out for Cambridge. Alice, our eleven year old daughter, now remembers with pleasure a night in the Garden House Hotel and the noise of running water. But I was not happy and for Ruth and me it was a sad journey. I had little idea of what I should do at Cambridge; I felt that teaching, rather than the research, might be my first priority. We were very fortunate that John Cockcroft had offered to rent us his house, no. 31 Sedley Taylor Road, and we moved in there. Next day I went to the Cavendish, talked to the (graduate) Secretary of the Department—'What are your problems?' 'Well, Prof, I did want to talk to you about the Cavendish cricket match, academic staff versus assistants'. After Bristol, it was a new world.

CHAPTER THIRTEEN

Cavendish Professor 1954–71

I moved to Cambridge in July 1954. I had some interesting letters of congratulation. For instance, Fred Seitz, the American scientist who shared my interests most closely, wrote from the University of Illinois in January 1954

> Dear Mott:
>
> For several weeks there has been a consistent report that you have accepted the appointment as director of the Cavendish Laboratory. Deepest congratulations. During our sojourn in Tokyo I made bets with our colleages that this would occur, since it was the most logical solution of the issue. Either my judgment is lucky or very profound.
>
> Undoubtedly, this means that you will be very busy during the summer and that we will catch only fleeting glimpses of you at the various meetings. In any case, we all hope that Bristol will manage to hold its level and remain the very fine center of research which you have made it.
>
> Sincere regards,
> Frederick Seitz

The Cavendish, unlike Bristol in 1945, was a going concern with little opportunity to start new things; on the research side it took me some time to find a role. However, I came there with two very strongly felt beliefs about teaching. One was the need for a reform of the natural sciences tripos. This was the course, with its examinations, to be taken by all students of experimental science. It was mandatory for a student to spend the first two years on three separate experimental subjects; they could for instance be physics, chemistry and geology. This led to the examination called Part I. Only in October of the third year could a student start to concentrate on physics and be introduced to quantum mechanics and all that followed from it. With the final examination in May, there was no time to digest this new knowledge; the tendency was to learn things by heart. I was determined to allow a serious approach to modern physics in the second year. Bragg had been an enthusiastic supporter of the old system, but I hardly knew anyone else in the Cavendish who was. I had a strong ally in Alex Todd, Professor of Organic Chemistry, later Lord Todd and President of the Royal Society. He told me that he was not sure whether Cambridge or (was it?) Aberystwyth had the worst teaching of chemistry in the country; in Cambridge this was a result of the present regulations. But the biologists and geologists were strongly against any change. They feared that if science students were not compelled to taste their subjects, their numbers would dwindle. To get the change was a long struggle; I remember a woman colleague in another department saying—off handedly 'Don't break your heart'. This was good advice—I was even foolish enough, in a letter to my father, to say I would have to resign and get another job if I couldn't get it through; he, most wisely, told me gently not to be silly. But after perhaps three years, suddenly everyone agreed. Then we changed things so that in the second year, a student could take physics, advanced physics and one other subject. For several years after that I lectured on quantum mechanics to a large class of second year students. I am not sure that these lectures were appreciated by everyone! We used to ask for written comments on all lectures, even those by the Professor. I was surprised to read on one occasion 'Professor Mott does not lack enthusiasm, but he does not communicate much.' This was because I used to stop in the middle of a lecture and ask for questions—'did they understand?' I guess the students who wrote in these terms would have liked the sort of lectures that they could copy down in detailed notes, and reproduce them in the examination. I was unrepentant; I thought they could use books for that.

My second mission in education arose from my belief that the Cambridge scholarship examination led to over-specialization in the schools. Since these examinations were in the hands of the colleges, it was more difficult to make a direct attack; I came back to it when I was Master of my college (see Chapter 15). But I remember making visits to some schools to see what they thought. One was Oundle; when I arrived there, they asked me 'What are *you* doing here?' That the holder of Rutherford's chair should be interested in sixth form teaching surprised them. I also went to Manchester Grammar School which was under Eric James, later Lord James and first Vice-Chancellor of York University. A letter to my father describes my reactions.

I found Eric James and thirty or forty of his staff and several bottles of sherry. I talked about specialisation and how it affected the universities and then there was a free for all. Obviously they have a magnificent staff; we had a vigorous and uninhibited discussion, and whatever Manchester Grammar School staff does in Manchester Grammar School is right. But what about Todmorden Grammar School? We talked about that too. James believes in the two 'elite' universities, the entrance scholarship and (doubtless) the elite schools. 'Take away the (Oxbridge) entrance scholarships and scholarship in the schools will go', he said.

I believe that the 'elite' universities fall down in their job in too many ways and the essential thing is to upgrade some civic universities so that Oxford and Cambridge have to fight for their position.

How do they fall down? The eight week term (modern universities have ten); and the absence of contact in colleges between dons and students. I had Mervyn Stockwood (then Vicar of the University Church) to a feast last night, and he was of the opinion that few dons made any attempt except the deans and chaplains. Reason—business, marriage, washing up . . . I think we did more in Bristol, organised in departments, than Cambridge really does with its colleges.

Certainly I felt that educational problems were my priority. I was chairman of the education committee of the Institute of Physics. At that time the Nuffield Foundation was funding a major attempt to reform and modernize science teaching in schools, to produce a course in physics, chemistry and biology at O-level which everyone in a grammar school could take. Later they produced a modernized A-level course too. I was asked to chair the Nuffield advisory committee on physical science. I had, of course, no experience of school teaching, but what skill I had in managing committees was amply needed. I particularly remember a meeting when the full-time organizer came up with a scheme for A-level in which sixth formers would be introduced to quantum mechanics and relativity. He argued that, while history students were told about the latest thinking of a Herbert Butterfield, it was difficult for science teachers to attract talented children when the curriculum stopped at 1900. But the three University people on the committee objected in the very strongest terms. 'Anything you teach about modern physics *before* the children have understood the earlier work will be at the level of BBC popular science,' they said. 'The schools must teach the foundations, and *we* will teach modern physics'. Obviously a compromise had to be hammered out. I hope we succeeded in doing it, under a new leader, Paul Black.

The main idea of the Nuffield course was 'learning by doing'—the student should be encouraged to discover the laws of physics by himself, rather than having them imparted to him through 'chalk and talk'. While I thought this was splendid, I realized that it was not the way I had learned science, and that I would have hated to have done it this way. To me the laws of physics are to be approached through mathematics, and their validity and beauty become clear when expressed in mathematical form. My hero was Clerk Maxwell rather than Faraday, whose laws of electromagnetic induction seemed so obviously right, and also Dirac who predicted his spinning electron theoretically, rather than the people who earlier showed by experiment that it was there.

I undertook two London jobs, at the instigation of Philip Morris, Bristol's Vice-Chancellor, who I am sure felt that education should have priority over research in the career of a Cavendish Professor. One was to be a member of the Committee on Education from 16–18 under the chairmanship of Geoffrey Crowther. Here I met some men who became life-long friends, particularly Alec Clegg, chief education officer of the West Riding in Yorkshire. He was very sympathetic to my views on over-specialization. He kept on quoting Mexborough Grammar school in the Yorkshire coalfield. The bright boys aimed at Oxbridge—but they paid for it by concentrating so much on their mathematics and physics that they missed other aspects of culture, which they might not find in their homes. Could the colleges not admit a year early, he asked, on the headmaster's recommendation and subject to a reasonable pass at A-level, so that the pressure was lifted from the sixth form? When I was master of my college we did try this in a small way.

My other involvement in education, also I am sure arranged by Philip Morris, was to join the Ministry of Education's committee on the Training and Supply of Teachers and to be chairman of the Committee on Supply. This I enjoyed. John Fulton, the first Vice-Chancellor of the new University of Sussex, was chairman of the main committee and believed strongly that teachers should have three years of education and one of training for the job. I did not; I believed that professional training could be as good an education as any. I remember always Eric Ashby's (later Lord Ashby, Vice-Chancellor of Belfast and afterwards Master of Clare College, Cambridge) praise of an education based on the technology of brewing,[10] which would include chemistry through the process of fermentation, botany through the cultivation of hops, architecture in the design of public houses and ethics in the consideration of over-indulgence. The Committee on Teacher Supply concerned itself with the shortage of mathematics teachers. With Hermann Bondi we recommended that university mathematics departments should concentrate not only on the honours man, the future research worker, but also on those taking less exacting courses, perhaps mixed with other subjects, that would attract more students who might opt for teaching. But whether this had any effect I do not know.

My involvement in school education continued till many years after my retirement, and I served for ten years from 1965 on the Education Committee of the Royal Society.

Other educational activities outside Cambridge included membership of the committee which planned Sussex University, and later that which planned Norwich. The main idea at Sussex was to get away from self-contained departments, in favour of 'schools' covering wider fields such as physical science or European studies. I doubt whether in the event this made much difference. I have been told that it led to too much committee work in maintaining liaison within the schools, so that a departmental structure gradually superimposed itself. During this time I was asked informally more than once if I would like to take on the Vice-Chancellorship of one of the new universities. I was tempted, but said no.

Also I was chairman for many years of the 'National Extension College'—a correspondence college in Cambridge started by Michael Young (later Lord Young) and Brian Jackson which was in many ways the forerunner of the Open University.

But my most interesting experience in the field of education was an approach made to me by Kurt Hahn. Hahn had been secretary to Prince Max of Baden at the end of the first world war, had created a remarkable school at Salem near the Lake of Constance and had left Germany after the Nazis came to power and founded Gordonstoun School in Scotland. I think he approached me because my citation in *Who's Who* included my having been President of the Modern Languages Association. I want to say that, of men I have known at some period of my life rather well, Kurt Hahn, Philip Morris and Niels Bohr have impressed me most as having that indefinable quality of greatness. (Perhaps I should add Patrick Blackett, at any rate the younger Blackett.) What Hahn wanted was the following. He had induced their chairman Sir Walter Benton-Jones to persuade the United Steel Company to provide the funds for 'Trevelyan Scholarships' for entrance into Oxford and Cambridge; sponsors had been found to sell the idea in Oxford; would I try in Cambridge?

The history of the Trevelyan scholarships has been told in a small book published by the Trust, before it was wound up. The idea as put to me by Hahn was that sixth-form education, concentrating on A-level, denied a boy or girl the opportunities to engage in other activities which demanded courage, perseverance and application, which were needed for his full development. So financially attractive scholarships were to be offered to young men and women who, as well as being excellent academically, would submit a written account of a project which demanded all these qualities. This idea agreed very much with what I had come to feel about sixth-form education, and I threw myself enthusiastically into the task of trying to gain acceptance for the scheme, particularly among college tutors. It was a sticky ride, and felt by some as an odd thing for the Cavendish professor to be up to. It was dangerously near demanding 'character' rather than 'intellect', an idea that I'm sure appealed to Benton-Jones, but which was of course anathema to me. I saw the scheme as a public recognition by Cambridge that, though prizing intellectual distinction before all things, we recognized that intellect developed best in an environment where other qualities can grow too. I was involved with these scholarships throughout, attending meetings in London, examining and so on. Also I got to know Hahn well, stayed with him twice at Salem and met many of his friends (including the Duke of Edinburgh).

As regards prizing intellectual distinction before all things, a curious incident occurred in Johannesburg, which I visited in 1962. As Master of Caius I was handsomely entertained by some Caians of the business community, to a dinner at a very British club, with Annigoni's portrait of the Queen on the wall, and it all began well. But they (and doubtless I) had plenty to drink and they began to abuse me and Cambridge. 'Let's have less of this Tripos,' they said, 'Why doesn't Caius admit more rugger-playing men of character, who will come out and make a career here and keep the Boers in their place.' Useless to explain about Kurt Hahn—I don't think I tried. Next day one of them—rather sheepish-looking after the previous night's expression of strong feeling—took me to see a gold mine, where a plane-load of bewildered-looking blacks flown in from Mozambique were learning to use a spade.

Of course my main job was, or should have been, the Cavendish. I did feel that much that I wanted to change needed an understanding of the administration and power structure in the university—so different from that of Bristol. 'That cold, dead place' (see also Chapter 12)— Philip Morris had called Cambridge, hardly justly, but perhaps he meant that there was no leadership at the top, and no-one to speak for it. I was not alone in being critical. Lennard-Jones, my predecessor at

Bristol, had moved to the University of North Staffordshire as Vice-Chancellor, and there died. I wrote his obituary for the Royal Society. In a letter to my father I say (10.4.55):

> I called on Lady Lennard-Jones yesterday—she has moved back to Cambridge. She showed me a lot of old correspondence—I was interested to find that, when he came back to Cambridge after the war with his experience of ministries, he longed for a policy-making body here, was exasperated by its absence and tried to form one by setting up an informal committee of science professors. It never became official and now no longer exists.
>
> He was also for ever trying to get the University to say how big it wanted to be and how many research students it ought to have—as though he were drawing up plans for the Armament Research Department. The poor old university can't do that sort of thing. And yet it should—so many steps, pushed by sectional interests—like the big expansion in oriental languages for instance or applied economics, have led to results that the university can't digest—big bodies of men that the colleges don't want as teaching fellows and who thus sit on the periphery of the university life. L-J was quite right that if the university is to do this sort of thing it does need a post-graduate college.

Feeling as I did that the problems I faced concerned the university, when quite early in my time in Cambridge I got the opportunity to become a member of the General Board, I accepted, and remained so for seven years. The General Board, a committee of twelve under the chairmanship of the Vice-Chancellor, is the body that makes decisions on most matters, such as which departments should get a new lecturer and which should not. The Vice-Chancellor is the head of a college and serves only two years, too short to acquire much influence. Twelve members were too many to decide details—so a 'Committee on Needs' existed to sort things out. I was a member of that for a time. Of course all decisions had to be submitted to the Regent House—that is the whole body of dons, and twelve signatures could force a vote, but this happened only for major issues—such as whether Lord Hailsham should receive an honorary doctorate. (The vote against, on the grounds that as Minister of Science, he had done nothing to halt the 'brain drain' to the USA, was not successful). I remember certain issues in which I became involved. Some departments, such as physics, were expanding because the number of students admitted by the colleges increased. So such departments applied for more staff and, funds being available, got them. But it was felt by some, particularly the Secretary-General

(Harold Taylor) that even if funds were available the total number of dons should not increase. One reason was that most university teaching officers wished to be college Fellows, felt themselves second class citizens if they were not, and the places in the colleges were limited. So if there was to be expansion in—for instance—physics, there had to be compensating contraction elsewhere. But the question was—where? If a scientific department was envisaged, the professor would fight to the death for his private empire. The General Board picked on a very small department, Colloid Science, from which the professor was retiring. But even so, they could not bring it off. The department found enough friends to come and vote against General Board tyranny. Compensating contraction seemed impossible.

Another strongly debated issue was whether medical professors should be paid more than other professors—as they were in most universities. I voted yes. Indeed, one issue after another of this kind came up, and showed me how the university worked.

Before I joined the General Board, two other heads of departments had written suggesting that a committee of Heads of Departments should be set up, since on the scientific side they were the people with responsibility. I took this up, remembering the Committee of Deans at Bristol. When Noel Annan, Provost of Kings, was on the Board, the Annan Committee examined the decision-making process in the University. In the event, what happened was the strengthening of an almost defunct body, the Council of the School of the Physical Sciences, which acquired a full-time secretary. This was a fairly important change in our administration, bringing together the Heads of Departments and giving them some collective power. I think it is something that I started, and others finished it off.

In Cambridge it was only the scientific departments that had heads; they were paid slightly more than the other professors, because of the responsibilities they had to undertake. In the arts faculties the chairmen of the Faculty Boards, who changed frequently, had to guide their boards to wise decisions. No 'schools' were set up for them.

As a member of Council of Senate I took part in the tripartite discussions with City and County on planning matters. Towards the end of my tenure of the Cavendish chair I was chairman of a committee which tried to formulate the University's attitude to local science-based industry, with Ian Nicol as secretary. We argued that some such development should be strongly encouraged, both for the sake of science and engineering students and for good relations with the city. The *Mott Report* was instrumental in enabling Trinity College to open its Science Park, to which the Master of that college paid generous tribute in the opening ceremony in the summer of 1975. The massive development of technological industry in Cambridge is now (1985) referred to as the 'Cambridge phenomenon'.

Turning now to the Cavendish as a research institution, my first problem was the Medical Research Council's Unit on molecular biology. This was Bragg's creation, engaged in some of the most important and successful work of the time. They occupied a good deal of space and wanted more. One option would have been—as always—to back talent—and to make it a major feature of the laboratory. I did not do that. Their research students were not qualified for ordinary physics jobs and their staff did not teach, particularly in the practical classes. I remember Francis Crick, one of the group and a future Nobel prizewinner, treating with contempt our suggestion that they should. I judged that they would develop into a bigger thing than we could cope with, and encouraged the General Board to find them a new site. (J. G. Crowther's book on the Cavendish says that I was very sad when they left, but this was not so). The space they left enabled me to build up electron microscopy, and develop the work on dislocations that I describe in Chapter 14.

Then there was Philip Bowden's unit. Philip was a Fellow of my college and an intimate friend. He had built up a group on surface science in the Department of Physical Chemistry, and the department was shortly to move to the new laboratory at Lensfield Road. Philip felt that his unit was not supported or properly backed by the head of his department, and this may well have been so. Anyhow the idea grew up between us that the unit should transfer to the Cavendish, taking over most of the space that Physical Chemistry would leave behind. This took a good deal of fixing, and was the sort of thing that membership of the General Board facilitated. I think myself that this is one of the best things I did for the Cavendish. Philip Bowden died in 1968, and his unit still retains a touching loyalty to his memory. His achievement was to create a unit, within a physics department, where chemists and physicists worked together. This is the way to do solid state science, and the unit in the Cavendish has lasted till today and is very highly regarded. Since my retirement I have chosen to have my office in this unit.

Until his death, Philip was a great pleasure to have in the laboratory. At the beginning his unit was more or less self-financing and the technical and most of the academic staff were supported by outside funds. He was naturally anxious to get them transferred to the Univer-

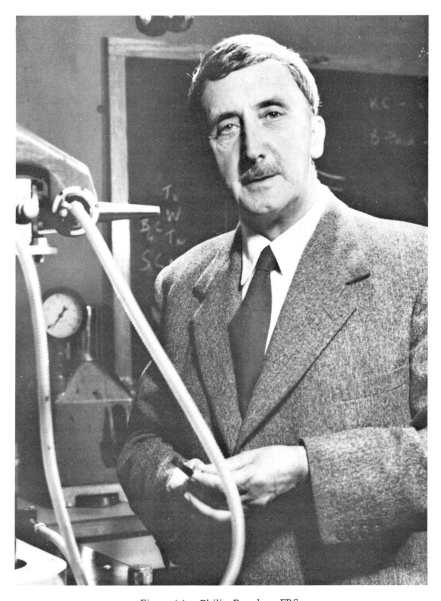

Figure 14. Philip Bowden, FRS.

sity, and now and then he came along to grumble that this was not going fast enough. 'Damn it,' he used to say, 'the laboratory has an international reputation' (he meant by this, his laboratory), 'and damn it, we only get . . .' I used to call these meetings 'Philip's damnits'. We remained excellent friends.

It seemed to me right to close down the old nuclear equipment, (two Cockcroft generators and the cyclotron) and to devote the considerable space and funds to other uses and to encourage a small high energy group to develop in co-operation with national centres and CERN. (CERN (Centre Européenne de Recherche Nucléaire) is the international centre for high energy physics at Geneva.) This worked out reasonably well. In nuclear physics the shadow of Rutherford was something that had to be lived with; it must have been much worse for Bragg. I remember in some journal (*New Scientist?*) a reference to 'the low reputation of the Cavendish'. I knew that to the writer this meant nuclear physics or perhaps just the absence of advanced equipment, but none the less it was hurtful. Then the old boys of Rutherford's day, when they saw what was going on, used perhaps unintentionally to express the feeling, 'What a pity'. Particle physicists still have an intuitive feeling that theirs is the only true pure physics. I reviewed (1977) Maurice Goldsmith's book on the new particle accelerator at CERN, and noticed that, when Mrs Thatcher brought the UK back into the project, the UK was said to be 'back in pure science'.

Another aim was to give the theorists in the mathematics faculty somewhere to sit, near the Cavendish. I was able to get a top floor to the Austin Wing on the New Museums site built for the purpose. However, with the departmentalization of the mathematics faculty, the creation of a Department of Applied Mathematics and Theoretical Physics, under George Batchelor, created problems. It appeared that Batchelor did not want his staff to sit under the patronage of the Cavendish. His concept of a unified mathematics department, and mine of theorists and experimentalists working together (as in Bristol), led to a period of friction. In the event one of Batchelor's people (a nuclear theorist) opted to transfer to the Cavendish, and on the whole those parts of theoretical physics which are closely related to the Cavendish are now carried out by Cavendish staff. However I think the problem of the relationship between mathematics and physics in Cambridge is likely to recur at any time.

Towards the end of my tenure the problem of more space for the Cavendish was beginning to become urgent. There were plans to develop the New Museums Site with a group of high buildings. This was strongly opposed, and in the end I was against it. I remember an experiment with tethered balloons, which showed that these buildings would show from the Backs, towering over the famous view from Queens Road to the Gibbs Building in Kings. The scheme was short lived, and it was widely—but not universally—felt that the only solution would be to go to west Cambridge, where the University owned a site rather more than a mile from the colleges. When this was finally agreed, the planning of the new building was undertaken by Brian Pippard, who, as I always expected, turned out to be my successor.

Naturally enough, finance of the department was a continual preoccupation. I remember organizing a week-end meeting for members of parliament and others under the title 'University Grants Committee; dirigiste or laisser-faire?' My thoughts about the problem are set out in the enclosed extract from the *Cambridge Review* (1955).

The Financing of Scientific Research

Reflections on the discussion in the Senate House on the General Board's report on research supported by outside bodies.
By N.F Mott

Cavendish Professor of Experimental Physics

This was a most interesting discussion to anyone concerned with the organisation of research. It revealed one fact that one might have guessed: the proportion, 12 per cent, of the expenditure of the University which is obtained from outside bodies is just what is normal for the other universities of this country, so the problem, if it is a problem, is by no means peculiar to Cambridge. Our University seems, however, to have been the first to investigate the situation, to publish its findings and to ask itself if it likes what has been found.

What then is the problem? The problem is the finance in universities of research in the natural and social sciences. Should it be financed by each university using the block grant received from the Treasury through the University Grants Committee, or should it be financed through other agencies, mainly those of the Government? Or, since research is now financed through both channels, are we satisfied with the present balance? Or are we afraid that the influence of these government agencies will be too great, and that it will distort the pattern of the research that a university such as Cambridge ought to have?

Before making up our minds, we might look at the kind of research that is actually done in our universities. A friend of mine just after the war was offered at the same time a chair in a university and the directorship of an industrial laboratory, and chose industry. He told me why. 'I'm a lazy man,' he said. 'In universities you have to think of your problems; in industry your're handed them on a plate.' Directors of industrial laboratories are rarely lazy, but apart from this he summed it up very well. In universities the choice of problem of every research worker experienced enough to choose ought to be made as a result of his own estimate of its value and importance to his science. Universities, too, are the proper place to do this sort of problem. At the same time it would be nonsense to pretend that the majority of these problems are unrelated to the immediate needs of the community, or that they ought to be. Medicine, Engineering and Economics would be sterile if they were not so related. Even in Physics, as the frontiers of knowledge move away from the atom into the inner recesses of the nucleus and to the nebulae of outer space, great fields of 'pure' research owe at least part of their appeal to the fact that their subject matters, such as metals and semi-conductors, are useful materials. There are no really sharp lines between pure research, applied research and technology. The task of a university is to break fresh ground in any of them, letting the talented worker follow his interests in the subjects of his choice, but avoiding only the kind of problem that is handed him on a plate.

The problem in the finance of university research is how to see that the talented man with a good idea gets ample support—and at the same time, since research assistants, technicians and so on are in desperately short supply, how to see that the man whose work has come to an unproductive stage doesn't get it. Support for pure research always has to be fought for; we have to ask, should our efforts be concentrated on the University, to urge larger block grants from the University Grants Committee, or on the other government agencies, the Department of Scientific and Industrial Research, the Agricultural Research Council and the Medical Research Council.

One can say at once that permanent jobs within the University, whether in teaching or research, are and must be the responsibility of the University, as must also be new buildings. But as regards equipment, assistance on temporary appointments and so on I am not so sure. Of course, no scientist, certainly no head of department, could do other than welcome increased university funds for equipment and research fellowships. But in the national interest I believe that it is even more important that more funds for university research should be made available to the Department of Scientific and Industrial Research and to the other agencies. The reason is that these agencies are often better able to judge who ought to have the money. After all, anyone who wants money for research has to persuade someone that he ought to have it. And anyone who has money to give must take advice on how to give it; the university will be advised by its professors, the Department of Scientific and Industrial Research at least partly by the same professors acting in another capacity. But it seems to me that the Department of Scientific and Industrial Research and these other agencies, drawing on the scientific opinion of the country as a whole, are better able to decide what ought to be supported and what ought to be stopped.

And this brings me back to the original question—do we fear the influence of government agencies in turning our efforts to the things that the government wants done? Certainly the Department of Scientific and Industrial Research shows no signs of wanting to exert pressure. Moreover I think that the question is wrongly put. The government is advised on these matters by the scientific community, largely by just the same people who seek to form policy on these matters within the Universities. The pressure such as it is comes more from industry and commerce who from time to time offer money for broad fields of work of interest to them. But this is not wholly unhealthy. Young research workers are very much subject to fashion in their choice of subject for research, and money offered from outside may turn a man to an interesting if less fashionable subject in which he can make his mark and which he would not otherwise have noticed. Moreover, industry can be very generous even to the purest research. Personally, under the conditions in this country, I think that outside support for research in our universities is to be welcomed, and that we need more of it.

My tenure of the Cavendish Chair included, of course, a period of student unrest. This did not affect Cambridge nearly as much as some other universities, but was not negligible; Lord Devlin reported on the student sit-in in February 1972 and its consequences (*Cambridge University Reporter*, Special Number 12, Vol. *CIII*, 1973). However most physics students were little involved; I remember the Vice-Chancellor at that time, Eric Ashby, remarking on how far we in the Cavendish seemed from what was his major preoccupation. But the fact that we accepted small amounts of money from the US armed services was discovered, and provoked some protest. We found this money particularly useful, because it was not tied to any particular project, so we were not deterred from accepting it. My most vivid memory, however, is of a long and friendly talk with two or three students about university decision making, ending with the remark with a rather sheepish smile—'but we ought not to be talking to you like this; it is a time of student unrest.'

CHAPTER FOURTEEN

Some Research up to 1960

During my first years as head of the Cavendish it was my job to get to know the various research groups. Radio-astronomy under Martin Ryle was all set for great success. (In 1974 Ryle and his collaborator Antony Hewish were to be awarded the Nobel prize for physics for their work on this subject). I was heavily involved in efforts to find a site for their observatory. We needed an open site not too far from Cambridge, for which planning permission could be obtained, far from motorways or houses where electrical appliances such as hairdriers emitted radiation, which would make the observation of radio noise from outer space impossible. Eventually the closing of the railway from Cambridge to Bedford made a site available. Ryle's observations gave support to the theory of an expanding universe, created by the 'big bang'. On the other hand Fred Hoyle, Plumian Professor of Astronomy, believed in the continuous creation of matter, so that the universe had no beginning. This led to an embarrassing experience for me. The King of Greece and his Queen, Frederica, who was an amateur of science, wrote to Hoyle asking if they could visit Ryle's radio-astronomy group. These two primadonnas were scarcely on speaking terms and they both poured out their resentments to me. I took the party round, and in the event all went well, but it was an exhausting day.

I had nothing scientific to contribute to radio-astronomy; the first research group with which I became involved was that run by a Hungarian emigré Egon Orowan, whose work had strongly attracted me when I was deciding whether to come to Cambridge. He soon left for America, but one of his Cambridge colleagues was another research worker from central Europe, Peter Hirsch, who later became Sir Peter, and held a chair at Oxford as well as the chairmanship of the UK Atomic Energy Authority, and continues to have a very distinguished career. He and his group were working with a newly developed instrument, the electron microscope, and gave the first definite evidence for the existence in solids of 'dislocations'. At that time I published several papers on the theory of dislocations, and, though I do not think they were of major importance, I had been involved with this concept since 1934, and will now explain what dislocations are and what we did about them.

A striking feature of metals is that, in general, pure metals are soft and easily bent, while alloys, that is to say the result of melting two metals together, are much harder. Our bronze age ancestors knew this. Bronze is a mixture of copper and tin, and the Homeric heroes would have been hard put to it to fight with copper weapons; these would have bent. We now know that the tin atom is much bigger than the copper atom, and the hardening resulting from the presence of a few per cent of some different atoms depends on this difference in size.

It is remarkable that no effort was made to explain this till the late 1930s, long after the discovery, for instance, of the neutron. Now in the 1980s, there are so many scientists around that everthing gets explored; in the 1930s, on the other hand, the exciting fields of the electron and the nucleus attracted nearly everyone. But in 1934 G. I. Taylor in Cambridge asked himself—how do the atoms in a solid move when a metal is deformed? He thought that one layer of atoms, in the small crystals of which metals are built up, could not slide over another, as a carpet would if pulled over the floor. It would be rather as if a ruck were made in the carpet, and this were carefully edged across. This ruck was called a dislocation, and figure 15 shows how he illustrated them in his paper. He thought that these dislocations would move very easily,

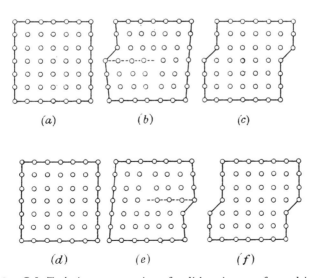

Figure 15.   G.I. Taylor's representation of a dislocation put forward in 1934.

if the material were stressed. A 'tough' material, bronze, would be one in which there was some obstacle to their motion. Just before the war, with a research student from Oxford, Frank Nabarro, I had a look at the problem of why certain aluminium alloys used in aircraft construction had an exceptional resistance to deformation, and came to the conclusion that very small inclusions of other materials could and did hold up the dislocations. Frank Nabarro, later Professor of Physics in Witwatersrand, (S. Africa) went on to make his career in this subject and published the standard book about it.

Research of this kind was in abeyance during the war; if a new alloy was needed, the skill of the metallurgists could produce it without thinking about the theory of dislocations. But after the war in Bristol I introduced my new colleagues to it, particularly Charles Frank and Jack Mitchell. Charles produced a beautiful theory of how they were generated (the Frank-Read source) and also a theory of how they affected crystal growth. Experimental evidence that he was right quickly came to light; Frank said, in a lecture at a summer school, that spiral markings on the surface of crystals should be present, and there was great excitement when someone at the back of the room said he had

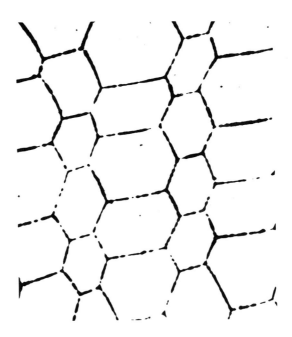

Figure 16. Network of dislocations in silver bromide crystal, made visible by deposition of silver (courtesy of J.W. Mitchell).

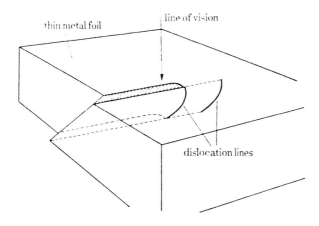

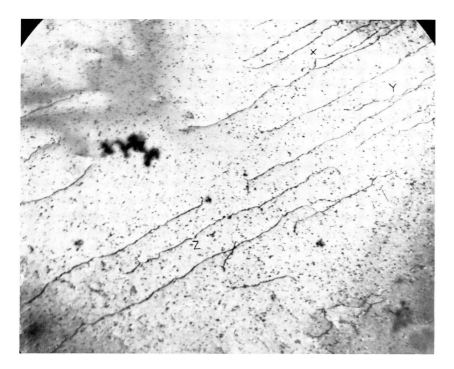

Figure 17.   (*a*)   Showing schematically the deformation of a metal foil under the electron miscroscope, with dislocations separating the region that has slipped from that which has not.
(*b*)   Cusps in dislocations in copper held up at points X, Y, Z. Courtesy of M.J. Whelan.

seen just this. Mitchell, on the other hand, had been fascinated by the theory of the photographic latent image which Ronald Gurney and I had produced before the war. The raw material of a photographic film, silver bromide, was known to decompose into silver and bromide when exposed to light, and Gurney and I had explained how this happened. But we did not know where the silver went, and Mitchell set himself to find out. By examining crystals of the material under the microscope, he saw a pattern of lines; an example is shown in Figure 16. His hypothesis was that they were dislocations. Figure 15 shows that there is more room for the silver to go near a dislocation, so the dislocation was 'decorated'—as we said—by the silver. Frank, for his part, managed to explain the form of these networks.

We had not, however, seen a moving dislocation, neither did we have any real evidence that they could move. Of course, once they were decorated with silver, we expected them to be stuck. The evidence came from the work of Peter Hirsch and his colleagues in my early days as Cavendish professor. Their electron microscope passed a beam of electrons through a thin metal foil, greatly magnified the image and showed it on a fluorescent screen. What they saw is shown in Figure 17, and, as Hirsch had expected, the dislocations showed up as the lines across the specimen. But what was exciting was that one could see them darting about, held up by impurities (Fig. 1(b)), doing all that one expected and a lot more. I well remember the day when some of Peter's young men came into my room and said 'Prof, come and see a moving dislocation'.

In 1957 I gave the Page Barbour lectures in the University of Virginia—lectures meant for a non-scientific audience. I chose this subject—so pictorial and easy to explain, and published them under the title *Atomic Physics and the Strength of Metals*. The book is now out of print, but it was a success and translated into several languages. 'Prof, come and see ...' impressed some students of European countries, where in those days such familiarities were unusual, and some Swedes sent me a sketch of the scene, myself as a stiff Swedish professor and the students in their little caps.

The discovery and observation of dislocations is one of the great stories of solid state physics. My contributions were small, except perhaps in the important task of directing young men towards the field.

CHAPTER FIFTEEN

Master of Caius 1959–66

In 1959 I was elected to be Master of my college, Gonville and Caius. In each Cambridge College the governing body consists of the Master and the Fellows. Most of the Fellows are teaching officers of the university, that is to say lecturers, readers and professors. They are elected to the fellowship by the College Council acting for the governing body and normally hold some post within the college which may be teaching or administrative. For this they receive a stipend in addition to their university salary. There are also privileges, such as a free dinner or lunch each day, a set of rooms in college for unmarried Fellows and—trivial but valued—the right to walk on the grass of the College courts. A few of the Fellows hold tenured positions wholly at the charge of the college; the Bursars who look after finance and property are often in this class. And every year a college such as Caius elects a number of research fellows by open competition, who typically hold the appointment for four years, and whose only duty is to do research. The day-to-day business is conducted by the College Council, which in my day met every other Friday at 2 and often sat until 7, but ultimate authority is with the whole body of Fellows who meet once a term under the chairmanship of the Master.

In most of the colleges, except Trinity and Churchill where the mastership is a Crown appointment, the Master is elected by the Fellows. He can be a university teaching officer, or a full-time salaried appointment; in Oxford he is always full time, but in Cambridge he is often a university professor as well. If he is, in Cambridge he will receive a handsome entertainment allowance and a free house. And—in the case of Caius—what a house! Part is medieval, some dates from the late sixteenth century, some including the splendid dining and drawing rooms from the eighteenth century and some is Victorian. There were five bathrooms. There was a hatch leading through into the college kitchens, so that any meal could be delivered, and waiters would be available too. The major rooms were partially furnished by the college.

The Master's duties were not entirely clear. He certainly did not have authority; any contentious issue was settled by the Fellows by vote. But he took the chair at all meetings and could have some influence. He attended the meetings of heads of other colleges. When his turn came, he was expected to serve as Vice-Chancellor of the university, for a period of two years, and chair all its main committees too, again without much authority.

It is the Cambridge colleges, not the university, that admit students, and to them the student pays a considerable part of the total fee. They also have endowments. Caius—or more properly Gonville and Caius College, was founded by Edmund Gonville, a Norfolk priest, in the fourteenth century as a small lodging house, and in the sixteenth refounded by John Caius (pronounced Keys), who studied at Padua and was physician to Edward VI, Mary and Elizabeth I. He built the present Caius Court, and the Gate of Honour, and gave the college new statutes. After refounding and endowing the college he was elected Master. But in the course of time his relationship with the Fellows deteriorated. At a time when extreme Protestantism was in vogue, he hankered after the old religion and was known to keep some church vessels and robes from the old days in his rooms. This caused some indignation, and once, when he was away in London, the Fellows took his treasures and, helped by like-minded heads of neighbouring colleges, made a bonfire of them in Caius Court, the court that he had himself built. John Caius, on returning and seeing what had happened, laid down the mastership, left Cambridge and died a few months later. He was brought back, given a splendid funeral and his tomb, inscribed FUI CAIUS, stands in the chapel.

When I came back to Cambridge in 1954 the Master was Sir James Chadwick, the very distinguished nuclear physicist. He announced his intentions of retiring from the mastership in 1958. I think the idea of succeeding him hardly occurred to me, till my colleague and friend Philip Bowden came into my office and put it to me. I asked whether he wouldn't like it; he said, it 'wasn't his kind of thing', but 'you could do it on your head'. Would I be a candidate?

First of all—why should I want it? Caius was a society in which I had never felt at home, and there was plenty to do at the Cavendish. But at that time I felt that I was unlikely to make a further major contribution to physics. My talent is piecing together experimental facts and interpreting them in terms of simple theory. This is what I did before the war in Bristol for metals and semiconductors, and after 1965 for non-crystalline materials, but I did not know about the latter then. What I did not want to do, or feel I could have any success in, was the fashionable attack on the so-called many-body theory using all the sophistication of modern mathematical techniques. This is the method of taking account of the obvious fact that in a metal there are huge numbers of electrons continually hitting each other, a fact that was of course always recognized but treated by a simple averaging process in the early work. I had never learned the appropriate mathematics and it seemed silly to start at the age of 53. My interests were increasingly in educational and administrative problems. It seemed to me that, if the mastership went well, I could resign my chair and perhaps serve for two years as Vice-Chancellor. In fact, some years later I told a colleague who was a member of the Council of Senate, the committee that chooses Vice-Chancellors (they are actually elected by the Regent House, that is the whole body of academic staff), that if they wanted me to be Vice-Chancellor, I would resign my chair. The colleague missed the meeting when the nomination was made; he told me afterwards that, if Council had known this, I would certainly have had the nomination. It was Caius's turn, anyhow.

Now I cannot write of the election or my experiences as Master, without describing those of my predecessor, Sir James Chadwick. Chadwick, a retiring, shy man with delicate health, was the discoverer of the neutron, the virtual director of Rutherford's Cavendish till he left for Liverpool in 1934, and the man on whom, during the later part of the war, co-operation between the USA and the UK in the Manhattan Project virtually depended. Margaret Gowing's books are his memorial; he proved himself a very great man. At the end of the war he returned to Liverpool and his beloved cyclotron where he formed an enthusiastic and devoted group, probably the best in Western Europe, but in 1947 he was asked to be Master of Caius, and accepted. In 1975 I wrote his obituary for *The Caian* (the college magazine) and tried to find out why he was willing to throw away all he had lived for. It was said that he wanted to pay his debt to his old college, which had given him a studentship to enable him to come from Manchester with Rutherford. I think he was the most loyal of men, and this was probably the heart of it. But it was a bad mistake for him.

Chadwick soon fell foul of a new generation of Fellows—not the first Master of Caius to have done so. Trying to find out what the issues were, while writing his obituary, proved hard. Should research Fellows qualify for FSSU (the university pension scheme)? How much of the endowment should be husbanded for buildings, how much spent on stipends and amenities? As always, who was most suitable for college jobs? But a group of Fellows took to meeting before College Council to

Figure 18. The author when Master of Caius representing the College at the patronal service of Lavenham church, a Caius living.

decide how they should vote. This was not how it had been done in the old days. The Professor of Music, who took all this lightly, called it 'The Peasants Revolt' because he thought the distinguished economist Peter Bauer, later Lord Bauer, had played some part in it (Bauer is German for peasant) (see reference 11) But I have no doubt that an 'anti-Chadwick' party organized itself round one of the tutors and that attacks on Chadwick were made in a wounding style. This really broke him and he retired prematurely, left Cambridge and, I believe, never entered the college again. This was a sad end to a fine career of public service.

This left the college a very divided place. There was a so-called progressive party containing some very interesting characters. Then there was the party of the old guard, very outraged by these goings on. The candidate of the progressive party could not get a majority. The old guard put up various candidates, and some 'progressive' Fellows told me they would 'leave the college' if the old guard won and 'ruled the college'. On the whole my sympathies were at first with the progressive party—the old guard represented what I had found uncongenial in

Caius, though I had hardly been aware of Chadwick's difficulties. When some of the progressives came into my camp I was flattered and decided to remain a candidate. It was a very long drawn out affair. Hugh Montefiore, Dean of the college and later Bishop of Birmingham was perhaps my most effective supporter, and told me that I could heal these divisions. So at length I was elected in 1959.

I remember our introduction to the Master's Lodge. The Bursar, E. P. Weller, took Ruth and me round, starting with the cellars. It was enormous, room after room, and needing much redecoration. This was lavishly done, and at length we moved in. We were instructed about our duties. One of Ruth's, according to the Dean, was to arrange the flowers in chapel. But it turned out that the other wives were very willing to help.

It seemed to me that at first I was successful as a Master and my friends told me that the old antagonisms had died down. I began to see the virtues of the 'old guard'. I found too that I could just cope with the college and Department. The Lodge was a splendid place for entertaining, and we gave many parties. A master is expected to attend some undergraduate functions; I went to the boat club's 'bumps suppers', relatively civilized and enjoyable, and the Rugby football dinner, rather less so. My presence, I am afraid, did not inspire the boat club to success and one year all of our boats were bumped four times. I received a letter from a former student, congratulating me on the performance—with the P.S. 'Sorry, I had turned the list upside down'.

My wife and I had pleasure in befriending the clergy who held college livings, mainly in Norfolk; we entertained them to lunch in the Lodge from time to time. Caius had a special relationship with Norfolk; up to the middle of the nineteenth century all masters had been Norfolk men. Together with the Master of Trinity Hall, I attended the enthronement of the Bishop of Norwich in Norwich cathedral. I remember the beauty and dignity of the occasion, and going afterwards for a drink in a large room with all the clergy, talking shop with the same absorption that one would have found among a crowd of physicists.

When I became Master, Caius was for men students only, and we had no women Fellows either; the Master and his wife were the only couple who could live in college. A Fellow could not bring in a woman guest to dinner (or to lunch) except on one night in the year, when at a special dinner each Fellow came with his wife (or bona fide fiancée); this dinner was known as Bishop Shaxton's Solace. But during my mastership opinion began to change. At one college meeting it was proposed that Fellows should be able to invite a lady to dinner in hall.

Some of the senior Fellows had doubts about this, and it was suggested that the privilege should be limited to academic ladies. From the chair I said that it would then be necessary to set up a committee to determine whether any lady invited was academic or not. Everyone laughed, and it was agreed that any lady could be invited. The next evening, a Newnham student was brought to sit next to the Master. I am glad to have been able to bring about this change in the college.

It was during my mastership that I received my knighthood, and this was celebrated suitably in the college.

There were quite important things to be attempted with a wealthy independent college. I was deeply interested in admissions policy. My ambition was that Caius should be known as the college that thought about admissions policy, not as part of the rat-race to get the best students, but as affecting school education too. I was able to do one or two little things, but not much. For instance, Alec Clegg, Chief Education Officer for the West Riding, had as I have stated earlier always talked of Mexborough Grammar School (in the Yorkshire coalfield), where the few boys who made it to Cambridge had to specialize so hard that they had no time for an introduction to literature or anything else. Could not a Cambridge college give a few bright boys a provisional acceptance after one year in the sixth, so as to relax the pressure? Caius' admissions committee was persuaded to examine the scheme, and go some way towards it. This was the sort of thing I hoped for Caius, but I doubt if I carried many of my colleagues along with me (see also p. 105).

During my mastership I was host to a Pugwash conference in my college. (Bertrand Russell and Albert Einstein had called for a meeting between Russian and Western scientists to discuss ways of avoiding the danger of nuclear war; the first meeting was held in 1957 in the village of Pugwash, in Nova Scotia, in the country house of an American business man, Cyrus P. Eaton. An annual meeting has been held ever since; the name 'Pugwash' has stuck.)

I had been interested in the control of nuclear weapons since 1945. After my return to Cambridge I had organized a discussion group on nuclear armaments. It met for several years at a colleague's house, and when a Pugwash meeting was to be held in England in the summer of 1962, I was able to offer the hall of Caius, free because our kitchens were out of action and a temporary dining Hall had been erected in Caius Court. I took the chair at this meeting. This was a fascinating experience. The opening session in the Hall of Caius was devoted to platitudes and the routine denunciation of the imperialists from the representatives of Eastern European countries; but then we separated into working parties on nuclear and on non-nuclear arms and so on. My working party was on non-nuclear disarmament in Europe; there were three Russians (including General Talensky), Kissinger and some others. I remember Kissinger, not yet famous, fencing with the Russians on their hostility to the Chinese, and the way they would retire into the next room to discuss how they could meet his points.

Ruth organized trips for the ladies, particularly the Russian ladies. Apparently they had been told that English ladies were very cold—but they thought better after they had met some of our friends. Ruth organized visits to 'a worker's house' (our daily help), Dr. Barnardo's homes, Ely cathedral, etc.—but eventually it was clear that what they really wanted to do was to go shopping, particularly at Marks and Spencer. We ended up drinking their vodka in our drawing room at Caius Master's Lodge, during a hilarious evening together.

I went to the next Pugwash meeting, Dubrovnik, in 1963—but after that felt that unless I became an arms control specialist, there was little I could contribute. In fact, before the Cambridge meeting (and the previous one in the USA), we did a lot of homework, under Blackett's leadership. To meet people like Admiral Buzzard and Alastair Buchan was a privilege, and for many years I was a member of the Institute of Strategic Studies. I became involved in these problems again much later, when we had left Cambridge.

I mention here a two-month trip round Africa that I made in 1962 as the Royal Society's Rutherford lecturer. Since Rutherford was a New Zealander, this lecture is given every two years in some Commonwealth country; I was lucky to be offered Africa. I spent three weeks in Uganda, visited Nairobi, Salisbury, Johannesburg, Nigeria and Ghana. My Rutherford lecture was on dislocations—a subject on which one can show lots of slides and give a comprehensible talk (see Chapter 13). The lecture in Kampala was a great success—buses came from all over the country, questions lasted an hour and the last one was, 'Will the professor tell us his life history please?' Indeed, Uganda, when I arrived three weeks after independence, was particularly enjoyable, visiting schools, talking about politics and Pugwash and having excellent discussions. Rhodesia was very different—contact with the black Africans was totally frustrating ('It's no use talking to us, we're helpless in international matters'). In Ghana I enjoyed meeting Conor Cruise O'Brien, just back from Katanga and Vice-Chancellor of the University, and spent an hour with the president of the country, Nkrumah, who, when we had talked about Pugwash and so on said, 'Let's be photographed.' Our photo in the Accra newspaper the next day made my driver wild with excitement, and he waved it whenever we had to pass a military road block, which was not uncommon.

The only way between East and West Africa was through South Africa. As that country was no longer part of the Commonwealth it was not part of my assignment from the Royal Society, but I spent a week in Johannesburg with my former pupil Frank Nabarro, who was professor at Witwatersrand. As Master of Caius, I was entertained by a group of former Caians whose reactions to my ideas on education are described in Chapter 12.

The end of my Caius mastership began when the Bursar, Weller, retired. I had come to have great confidence in his judgement. The College Council did not like my suggestions for a successor. At a meeting discussing it, a quite junior Fellow suddenly said that he had written to a former Fellow, who had been prominent in opposition to Chadwick, asking if he was willing to be considered, and that he had said he was. I was horrified; not that I had anything against the man personally, in fact I rather liked him, but I knew that he was at the heart of the group that had split the college in Chadwick's time, and that no more divisive move could be imagined than to try to bring him back. I consulted Bowden, who said, 'Sheer wickedness'; another senior Fellow said, 'Will these bastards never stop it?' I was able to show there was strong opposition to this move, and to prevent further negotiations with this particular man, but from then on there was no pretence that I was Master of a united college. It was, in the recollection of some, rather an unpleasant place. I made a lot of mistakes (I now think) in trying to see if a 'master's party' existed—something I am sure a Master should never do. Eventually it was clear that it didn't—not as a majority anyhow, and it was an enormous relief to resign in 1965, at the age of sixty. I wished I had done it earlier—but a mastership involves one's wife and family and resignation means moving house amid considerable publicity, an article in the gossip column of the Sunday Telegraph and so on. It was no joke for either of us.

One ex-Fellow, who had done much to persuade me to stand wrote to me, about my election seven years earlier, 'It was essential to introduce to that neurotic and twisted society a Master whose friendliness and largeness of outlook could bring into it *some* humanity. And this Ruth and you have done ...' and from a senior Fellow retired from his university position, 'how much I fear the result of a victory for that element of our Society that brought about the retirement of a former Master.'

My mistake was to assume that a mastership was like running a university physics department. There, both in Cambridge and Bristol, I tried to consult everyone and then made my own decisions. The consultations were informal, but on the whole people trusted me; someone has to decide, they felt, and better Mott than an elected committee. But Caius was the other way round. The decisions *were* made by the College Council. Any attempt to persuade them beforehand would have been resented, and if the decision went against the Master's wishes or feeling of what was right, so much the worse for him.

In 1975 I undertook to write an obituary of Sir James Chadwick for the College magazine—and the then Master of Caius (Joseph Needham) saw this in proof, and we talked about the matter. According to my recollections, talking of his own mastership, he said he had felt it the duty of the Master to accept in all matters the collective wisdom of the College Council, but found it none the less a never-ending struggle to procure peace, because the younger generation, with their tendency for intrigue and propaganda, caused divergencies to start all over again. And though individually he found the Fellows to be always amiable, in any meeting they were liable to say and do things collectively which could fill him with anger and mortification. It seems that three successive Masters have felt this, and though I have no direct experience of other colleges, I feel that Caius had at that time a flavour all its own. My theory—obviously only a theory—is that certain tactics—the forming of a small 'progressive' cabal, which fixed issues before college meetings—may have achieved some desirable reforms but at the same time led to a tradition of conflict, almost like militant students versus the administrations of later days. This produced an atmosphere of unnecessary antagonisms which lasted some time. But I think it has quite gone now (1985).

C. P. Snow's novel *The Masters* records the intense emotions that can accompany the election of a college master. Looking back on both my election and my resignation, it seems like one of Snow's novels; we used to say that Snow 'didn't know the half of it'. And yet the issues were trivial—personalities rather than principles. I can only say that—many years later—I enormously enjoy the company in college and have a great affection for it. I believe the college now to be a happy and harmonious community.

CHAPTER SIXTEEN

Non-crystalline Semiconductors

I was fortunate that, about the time that I left the mastership at Caius and wished to devote more time to physics, interest was developing in non-crystalline semiconductors. This was a new branch of solid state physics, in which it was possible to lay the foundations without unduly sophisticated mathematics, and to carry out experiments with relatively cheap equipment. The position was similar to that in crystalline semiconductors in the 1930s. From 1965 onwards research in this field was my main activity, as it could not have been had I remained a college Master, and for this work I shared a Nobel prize for physics in 1977. I will next describe the nature of this work, how I came to be involved in it and what it achieved.

First I must go back to my early work on metal–insulator transitions. I have written earlier (Chapter 3) how, at a lecture course which I attended as an undergraduate, I was intrigued by the problem of why metals contain free electrons, the movement of which constitutes an electric current, whereas in insulators the electrons are 'stuck'. The answer to this problem was obtained very soon after the formulation of quantum mechanics by the work of Felix Bloch[12] and Alan Wilson[13]. It depends essentially on the wave-like properties of the electron; if one wants to know where an electron is and where it is going, one has to postulate a wave (the wave function, denoted by ψ, which is a complex number) and suppose that $|\psi|^2$ denotes the chance that an electron will be found at a given place. Now, within a crystal, in which of course the atoms are arranged in a perfectly regular way, a wave can propagate without being scattered; this shows why pure metals at low temperatures are such good conductors. An electric field drags the electrons along, and they go a long way before being stopped. Electrical resistance arises either as the temperature rises, or because the metal contains impurities; the electrons are scattered more often, and so drift along more slowly. However, as the Braggs showed in their Nobel prize-winning work with X-rays, for certain wavelengths a wave cannot propagate in a crystal but is reflected. So the same must be true for a beam of electrons. Another principle, the exclusion principle, says that in a solid no two electrons can be described by waves with exactly the same wavelengths moving in the same direction. Electrons are paired, one moving in just the opposite direction to another, and if the shortest wavelength present coincides, as it often does, with the Bragg wavelength, the material is a non-metal. The electrons are not 'stuck', but exactly as many are moving in one direction as in the opposite direction. This rather unexpected way of thinking had been accepted widely before 1933, when I became involved in the subject, and is now the theoretical basis of silicon technology.

In 1949 I published a paper asking whether certain kinds of non-metallic materials would become metals under sufficient pressure. My conclusion that they would was not surprising; what was new was my demonstration that all the electrons would become free at once, not just a few of them. This kind of transition from non-metallic to metallic behaviour has been called in the scientific literature the 'Mott transition'. The way I attempted to prove it has been replaced by better methods[14], but the conclusion was correct. The question was—how to prove it experimentally. For most solids the pressures needed were not obtainable. However the semiconductors silicon and germanium now used in technology had been extensively investigated during the war; and it was known that, to make them conduct well, traces of another element such as phosphorus had to be added, which contained one more valence electron than did these elements, namely five instead of four. Of these, four would be taken up in bonds with the surrounding silicon atoms, but one was very loosely bound. It behaved like a greatly swollen hydrogen atom, and the electron could easily escape, heat at room temperature being sufficient to free it. Experiments at much lower temperatures showed that for low concentrations of phosphorus the heat motion was not enough to free the electron, so that the conductivity dropped to zero, as the temperature was lowered. On the other hand above a certain concentration it did not; the electrons were free. The critical concentration was about 3×10^{18} phosphorus atoms per cm^3, which means one atom of phosphorus in 30 000 silicon atoms. So above this concentration the electrons are free, as in a metal; below it, they are not.

Here then was a method of investigating the metal–insulator transition but there was one difficulty. My theories had been for a regular array of atoms as in a crystal; the positions of the phosphorus atom were random. Would this make any difference to my prediction, that all the electrons would suddenly become free as the concentration of phosphorus increased?

The first major paper on what happened to electrons in atoms in random positions was published in 1958 by P.W. (Phil) Anderson of the US Bell Laboratories. For several years Phil was a visiting professor at the Cavendish, spending half his time there and half at Bell Laboratories. In 1977 he shared the Nobel prize with me and one other, partly for this paper. It was entitled *Absence of diffusion in certain random lattices*; it was by no means easy reading, and Phil himself described it as 'Often quoted but rarely read'. It was often disbelieved; a distinguished American theorist spending a sabbatical year at the Cavendish told me that he neither understood nor believed it. But what the paper proved was that, given enough disorder in the arrangement of the atoms, electrons *could* be stuck in solids; I knew this was right, because I knew that for a low concentration of phosphorus atoms in germanium their electrons *were* stuck. This might of course have been simply a 'Mott transition'—but the latter was predicted only if every phosphorus atom had an electron; in so-called compensated† materials this was not so. Anderson localization, it seemed to me, was experimentally proved. In 1961, with my student Twose, I wrote an extensive review of the matter[15].

Since then experimental and theoretical work on this problem have both developed enormously. I have written perhaps 20 papers about it, several with my friend Moshe Kaveh from Bar Ilan University in Israel, who spent much time in the Cavendish and whom I visited in 1982 in Israel (they gave me an honorary doctorate). Work at very low temperatures at Bell Laboratories in 1980 showed that some of my early conclusions were wrong, as did theoretical work by Anderson and co-workers. At the time of writing the subject is still developing, with work in the USA, Russia, Japan and many Western European countries.

This problem, then, of the metal–insulator transition in silicon doped with phosphorus, together with Anderson's paper, introduced me to the problem of disorder. Another influence was the elegant theory of the electrical resistance of liquid metals, published by John Ziman in 1962. Ziman was the theorist in the Cavendish at that time; later he went to my old chair in Bristol. Atoms in liquid (unlike a

---

† Compensated material contained some boron, with three outer electrons as well as phosphorus. Each of these atoms takes an electron from a phosphorus, to form the four bonds with neighbours. So some phosphorus atoms had their outer electron, while some did not.

crystalline solid) are disordered; in a paper in 1966 I tried to extend his ideas to certain special cases. Another influence was the work of B.T. Kolomiets in Leningrad. His group had been investigating certain glassy semiconductors since the early 1960s; these were the so-called chalcogenide glasses of which a typical composition is As_2Te_3. They established the remarkable fact that these materials, unlike the crystals, did *not* suffer a great increase in conductivity through the addition of impurities. I gave an explanation of this in 1969 (the 8-N rule), still generally accepted.

In 1967 I attended a conference on metal–insulator transitions in San Francisco. Here I was introduced to Stan Ovshinsky, who gave a splendid party for me and some other participants in the city's Chinatown. Stan, a non-graduate but with an instinct for science, had started a small firm, Energy Conversion Devices (ECD) in a suburb of Detroit, to exploit the potential of non-crystalline materials in electronics. He persuaded me to visit his laboratory on the way home. He had developed an electrical switch, using a thin film of a chalcogenide glass with two electrodes. The giants of the industry pooh-poohed his achievement, and said there was nothing new. But I thought there was, as did other outstanding American physicists. Indeed, I published two papers trying to explain how it worked[16, 17].

Since then I have visited ECD many times. I enjoy the drive from the airport with Stan's loyal and intelligent driver (Mr. Max Powell who has worked for the company since 1960 and is in charge of liaison with visitors), the polyglot and enthusiastic team of scientists there, the hospitality (my favourite malt whisky) at Stan's house. Stan's dinner parties are a feature at all conferences on the subject, as are the meetings he calls from time to time at ECD.

All those influences, then, led me to look at the evidence we had about non-crystalline, that is glassy, semiconductors and to see whether I could piece together the laws which governed their behaviour. I soon found that new concepts were involved—'mobility edge', '8-N rule', 'minimum metallic conductivity', which aroused a lot of interest.

The first conference on this subject was in Prague in 1965, but this was before I became involved. It was however at first an Eastern European subject, thanks mainly to the Leningrad school, and the next meeting was in Bucharest in 1967. I intended to go, but just then my father died, shortly before his ninetieth birthday, so I could not make the journey; I visited Professor Grigorivici, another early contributor to the subject, in that city later. This was the year (1968) de Gaulle and his wife visited the city. I saw them driving up the main street. But he had to go back prematurely because of the student unrest in Paris. In 1969 I organized the third of these meetings in Cambridge. This was a success. I remember the very effective way my secretary Shirley Fieldhouse coped with lodgings, finance and every emergency. Many Czechs wanted to come, but could not get finance; a grant from Robert Maxwell of Pergamon Press, himself originally Czech, eased this. Several Russians came, but not Kolomiets, who was never able to attend conferences in the West. On the scientific side a highlight was the description by Walter Spear of the excellent photoconducting properties of films of amorphous silicon, deposited from a glow discharge in silane (SiH_4).

Further conferences were held in Ann Arbor, Edinburgh, Garmisch in Germany, Leningrad, Harvard, Grenoble, Tokyo and Rome, with the silicon solar cells becoming more and more the centre of interest.

Spear and his group in the university of Dundee later made the very important observation that those films—unlike Kolomiets's chalcogenides—*could* be doped. They could therefore be used for photovoltaics, that is films that can convert sunlight into electricity. This led to a world-wide interest in the material; Ovshinsky's firm, and the Japanese, took it up in a big way. Most readers will have used a pocket calculator powered by a solar cell; the optimists believed that solar power stations are just round the corner. If so it will be good to know that one outcome of this research should be wholly beneficial. I have remained in close contact with Spear's group, making the journey to Dundee once or twice a year. One of his main experimental tools is the 'field effect'—which my father tried to develop in 1902.† For these materials, a strong electric field perpendicular to the surface can change

---

† The electron was discovered about 1900, as a particle, which carried electricity when a current flowed through a gas at a low pressure, a phenomenon which had been known for some time. Since a beam of electrons constituted an electrical current, it was deflected in the presence of a magnet; that was how Thomson measured an important property of the electron, the ratio of the charge to the mass. It was known, too, that the electrical current in a metal wire was carried by electrons; the German physicist Drude, at the very beginning of the century, supposing that both current and heat were carried by the same kind of particle, showed that its ratio of charge to mass was just the same as for the electrons in a gas. So J. J. Thomson suggested to my father that he should make a thin film of some metal, and apply an electric field perpendicular to it, which would empty the surface area of some of its electrons and, hopefully, make the conductivity smaller in the plane of the film. But no effect was to be found. We know now that the density of electrons in a metal is much too big for the expected effect to be observable; for semiconductors, in which of course the density of electrons is much smaller, a similar effect is found and indeed used. It is now called 'the field effect'.

the density of electrons and make the film conduct better. In this way he could determine how much space for electrons was available, in technical terms the 'density of states'.

Another group with which I had the happiest collaboration was that of Josef Stuke in Marburg (West Germany). I first met him at a conference in Sheffield in 1968, and visited him often. His experimental work, particularly on electron spin resonance, proved very important. He organized the conference in Garmisch in 1975. The Volkswagen Foundation gave us a grant jointly to facilitate co-operation between our two laboratories, which enabled our students and ourselves to make as many visits as we needed, and paid too for some technical staff. It happened in this way. When we were planning a new building for the Cavendish (see Chapter 13), I tried to get private funds to build it, both from the Rockefeller Foundation and from Volkswagen (Kurt Hahn introduced me to them). They were not willing to contribute towards the many millions needed for a new Cavendish, but were attracted by the idea of helping a German and British university to work together. This help towards our relationship was very useful.

As I shall relate in the next Chapter, I learned of my Nobel prize when visiting Marburg. Later I came back to receive an honorary doctorate, and again to make a speech in his praise on the occasion of Professor Stuke's retirement.

My work in this field was not interrupted by my retirement in 1971 and even now, in 1985, I remain in close contact with many of the groups involved. I feel that I have groups strongly influenced by my work in most of the industrial countries of the world and keep in touch with them by correspondance. Also, I have written two books about it. *Metal–Insulator Transitions* published in 1974 by Taylor & Francis and *Electronic Processes in Non-Crystalline Materials*, in collaboration with E. A. Davis, published by Oxford in 1971; we have a second edition, extensively rewritten, published in 1979. Ted Davis was my collaborator for many years in Cambridge and is now professor of experimental physics in Leicester University.

CHAPTER SEVENTEEN

The Nobel Prize for Physics 1977

The Nobel prizes were founded under the terms of the will of Alfred Nobel, a Swedish manufacturer of explosives. They were first awarded in 1901, with prizes for physics, chemistry, medicine, literature and peace; in 1968 the Bank of Sweden put funds at the disposal of the Nobel Foundation for a prize in economics. Prizewinners and others can make proposals; the decisions are the responsibility of a committee of Swedish scientists appointed by the Foundation. The announcement is made in the autumn; the prizes are awarded in December, and, excepting the prize for peace, in Stockholm.

My three immediate predecessors in the Cavendish chair, J. J. Thomson, Rutherford and W. L. Bragg, had all been awarded the prize. (Rutherford's was for chemistry, because his work on sorting out the radioactive elements was classified in that way. In view of Rutherford's classification of science into physics and stamp collecting, this gave rise to some jokes.) I really did not expect to follow them, because the prize was usually given for one startling discovery, such as a new fundamental particle (Cecil Powell at Bristol), a basic new theory like quantum mechanics (Heisenberg and Schrödinger) or a very important device such as the transistor. I did not think I had done anything of this kind. Certainly I was aware that I had introduced some basic new ideas into our understanding of amorphous conductors, but many of these were not accepted by everyone—and one, the 'minimum metallic conductivity' did in fact turn out to be wrong. Before about 1975 I hardly thought about it.

In that year I was invited to go to Linköping in Sweden to receive the degree of Doctor of Technology, Honoris Causa. My host was the professor of physics, Karl Berggren, who had worked with me in Cambridge and who is a close personal friend. I remember at the ceremony the young Swedish king, and the long orations in Swedish in which the only word I could recognize was 'Linköping', repeated every minute or so. In some casual conversation the Nobel prize was mentioned and Karl, who must have been in the know, said 'of course you are high on the list'. This was hardly discreet and that year and the next I must confess to a slight feeling of disappointment in the autumn when my name did not appear.

In 1977 in early October Ruth and I went to Tübingen, because Cambridge had asked me to represent them at the five hundredth anniversary of the University. This was an extremely interesting occasion, with addresses in the cathedral (the Stiftskirche) from the president of the republic (Herr Scheel), Hans Küng and others, all against the background of the Bader-Meinhof terrorist gang, and the kidnapping of an industrialist who was afterwards murdered. The speeches asked whether the universities were responsible for the alienation of a generation of students.

There was tight security—given the presence of Herr Scheel. Also a friendly confusion. All the delegates brought Latin scripts, which we expected to read out, and I had even informed myself on how Germans pronounce Latin. But in the event there was not nearly time for this. Oxford, Cambridge, Paris, Montpellier, Bologna and many others agreed to present their scripts in silence, to make time for those who had travelled long distances from the developing world.

The next day we went to Marburg to see Professor Stuke and his colleagues, many of whom we knew well, though Ruth had never been there before. My work with him was mentioned in the last chapter. While Ruth was entertained by the wife of one of my colleagues, I went with two or three of them to lunch, as I had often done before, in a little restaurant *Die Sonne* in the old town. While we were lunching the telephone rang, and it was an English journalist who had traced me there through the laboratory secretary in Cambridge and then through Frau Stuke. 'What are you going to spend your Nobel prize money on?' he asked. 'First I've heard of it', I said. They explained about my sharing the prize with P. W. Anderson and J. H. van Vleck and I was delighted, especially to share with Phil Anderson; my work owed so much to him that I would not have been happy otherwise. Then at the lunch table Josef Stuke expressed his delight and said he was one who had proposed me. I felt dazed, asked to walk down alone to the Gothic Elizabethskirche to clear my mind and then back to the physics laboratory, where first the university and then the city representatives came in to give me congratulations; the medal of the city of Marburg sits on my desk to this day.

The next day we had to go. *Die Sonne* implored us to come to lunch at their expense, which we did; it was a sumptuous meal in a private room. Stuke rushed down to the travel agent to change our tickets to first class, and we journeyed on a golden autumn afternoon down the Rhine to Hook of Holland, to arrive after a mercifully smooth crossing to find a magnum of champagne (from Ovshinsky) on our doorstep and a torrent of letters and telegrams which continued for weeks. They were the best part of it. Prince Philip, Chancellor of the University of Cambridge and Shirley Williams, Secretary of State for Education and Science, wrote congratulating me on my service to both. I reproduce an extract from that from the President of Tübingen university where I had been a few days before, and Sister M. M. Hurst who had worked with me on the Nuffield Physics project. There follow extracts from letters from various friends who are physicists, and finally one from a colleague in Caius and his wife, Germanists, not scientists.

Es gereicht der Universität Tübingen zur Ehre, mit Ihnen nicht nur einen charmanten Gast, sondern auch einen hoch-angesehenen Physiker zu ihrem 500-jährigen Jubiläum beherbergt zu haben. Die Eberhard-Karls-Universität gratuliert Ihnen zum Nobelpreis 1977.
Mit freundlichen Grüssen (A. Theis)

Dear Sir Nevill,
You will be inundated with congratulations on the honour that has been done you, but I cannot forbear to add mine to them and tell you how thrilled I am.

I think it is particularly remarkable that you have, while continuing this high-level research, been able to give so much time, energy and high competence to the sphere of education, and I, for one, am personally very grateful.

Thank you for all you have done and the happiness that this Nobel Prize brings to your many friends. May the Lord continue to bless and prosper the work of your hands.

Yours sincerely,
Maureen Hurst
(Sister St. Joan of Arc, Nuffield Physics)

Dear Nevill,
I was in Grenoble when the news broke, and I have never known an award produce so much spontaneous and unalloyed pleasure. I doubt whether the Physics award has ever been as widely approved as this year—in both geographical and subject senses—since I doubt if any recent holder has been so deeply liked, admired and respected in so many places.

(Bryan Coles)

Dear Nevill,
The news of the Nobel award reached us when we were both at the Jahresversammlung of the Leopoldina at Halle. The awards to you and Prigogine were announced to the assembly by President Bethge and were received with tremendous applause. I myself must say that I was never before touched by a Nobel award in physics as much as by yours, not only because I feel that it was extremely well deserved but also since you have influenced my own career more than anyone else. Not only have I learnt from you how to combine theoretical and experimental physics efficiently but also that it is feasible and, in fact, very rewarding to be responsible for a large scientific institution and to do simultaneously 'one's own physics'. Had the compatibility of these two seemingly conflicting activities not been demonstrated so convincingly by you, my entire career would presumably have taken a different course.

(Alfred Seeger, Stuttgart)

Dear Nevill,
I find it almost impossible to express my delight, excitement and deep emotion when the news came through. You will of course be receiving scores of formal congratulations but may I just add my own very personal comment?

Alas, I was not sufficiently fortunate to have worked directly with you but your influence on the way I try and do things has been so profound; your kindness to me at all sorts of critical points in my life as a scientist is impossible to estimate. And of course, I am one of many many people you have helped in that way. The formal recognition of your work—so right and proper—fills me and many like me with extreme pleasure and delight.

Thank you for everything, and with jubilation I say CONGRATULATIONS!

(John Enderby, who holds my old chair at Bristol)

Dear Nevill,
I have a vivid memory of a remark you made on the occasion of your knighthood: that it was worth it, if for no other reason, than because of the evident pleasure it gave to your friends. What has now happened is in a more exalted region, of course; not, however, in the delight your distinction has given to those who hold you in high regard and affection. Please accept our heart-felt congratulations and good wishes: ad multos annos!

We hope things went off quite smoothly in Tübingen; and

how nice that you heard the good news in Marburg, a place well known to you and loved of old.

Warm greetings to you and to Ruth from us both.

(Freddy and Elizabeth Stopp
Gonville and Caius College)

In December we went to get the prize. Ruth and I flew to Gothenburg where I gave a lecture, then a whirlwind of a week of receptions, dinners and lectures in Stockholm, the same as all other prize-winners must have experienced. A car and a guide is attached to each laureate. We had a specially arranged tour around the Stockholm collection of Anglo-Saxon coins, the result of Danegeld paid by the Saxon kings. On the way back we passed through Linköping and met the students there. And we visited a married niece in Hamburg, and arrived home both as exhausted as we have ever been.

What difference does a Nobel prize make? To me, in my retirement, very little. I have had no inclination, having got there, to give up physics. There is a large increase in requests to add my signature to some appeal or other. Usually, I say no—not believing that the prize

Figure 19.   Philip Anderson, Nevill Mott, Ruth Mott, Joyce Anderson, Stockholm, December 1977.

should be used to add prestige to such things. Outside the scientific world, people wonder what it means; I have more than once found it assumed that the prize was for peace.

Perhaps the chief pleasure is the obvious satisfaction it gave friends and colleagues. Amorphous semiconductors was a new subject, and some—perhaps particularly Stan Ovshinsky—felt that it had made it respectable. Anyhow he gave a lunch to 200 Detroit industrialists and others in my honour, the next time I visited his firm Energy Conversion Devices in Detroit.

Finally, though some of my friends say I was unlucky not to get it earlier, I know how lucky I am to have it at all. For the giants, Rutherford, Einstein, Bohr, the prize was a certainty, but I do not judge myself in that class. I believe there was a rival name, or group of names, and the Nobel committee decided on us only on the very day it was announced, and I know of several colleagues whom I feel to be at least as worthy of the prize as I am.

CHAPTER EIGHTEEN

Honorary Degrees

I have quite a number of these—Louvain, Grenoble, Poitiers, Paris, Bristol, Ottawa, London, Reading, Liverpool, Sheffield, Oxford, Sussex, Essex, East Anglia, Warwick, Lancaster, Heriot-Watt, St. Andrews, William and Mary (USA), Stuttgart, Marburg, Linköping (Sweden), Bordeaux, Bar-Ilan (Israel), Lille, Lisbon, Rome and also the Fellowships of Imperial College and UMIST (Manchester). Most were very pleasant occasions, where the recipient stands on a platform and listens to an oration, usually flattering. One is reproduced below (Essex, 1978):

Professor Sir Nevill Mott M.A.,D.Sc.,F.R.S.

My Lord and Chancellor, the Senate had resolved that the degree of Doctor of the University be conferred on Professor Sir Nevill Mott.

In April 1928 a committee of the Department of Scientific and Industrial Research met to consider whether or not it would support a promising research student. The investigations of one of my colleagues reveal that the minutes of that meeting read in part as follows: 'The Committee considered an application ... that Mr N.F. Mott should be allowed to spend a large part of the academic year 1928–29 at Göttingen and Copenhagen working under ... Professors Bohr and Born respectively ... In view of Mr Mott's exceptional capacity and promise, and of the nature of his studies, the application should be granted.' Committees, as we all know, can never get everything right: as it happends, it was Professor Born who worked at Göttingen, and Professor Bohr who worked at Copenhagen. But in identifying N.F. Mott's exceptional capacity and promise, the committee was most certainly on to a good thing. Sir Nevill (as he has been since 1962) stands here this afternoon, to our great delight, with fifty years of scientific publications and more than forty as a Fellow of the Royal Society behind him, and the possessor of unrivalled eminence in British physics.

Sir Nevill's academic career has been as distinguished as it is possible to be: at the age of 28, a fact eloquent enough in itself, he became Professor of Theoretical Physics at Bristol, where he stayed until appointed to the Cavendish Chair at Cambridge in 1954; this he held until 1971. During this period he spent seven years as the affectionately-regarded Master of Gonville and Caius College. If the award of honours and distinctions is a measure of eminence, as to some extent it must be, then one wonders whether it is really possible to become very much more eminent than Sir Nevill. Leaving out of the account altogether a whole host of honorary doctorates of which this is merely the latest, there is the Hughes Medal of the Royal Society back in 1941; the Royal Medal twelve years later; and, within the past decade, the Grande Médaille de la Societé Française de Métallurgie, the Copley Medal and the Faraday Medal. It seems hardly surprising that one of Sir Nevill's reputed side-interests is numismatics. Last year, as most people will know, Sir Nevill received the ultimate accolade when he was awarded, with Anderson and Van Vleck, the Nobel Prize for Physics. That possibly completes the collection, though a betting man would be ill-advised to put money on it; the particular work for which Sir Nevill was awarded his Nobel Prize was done almost entirely *after* he retired. Did I say 'retired'? The word hardly seems to have much meaning in this case, although I do understand that he and Lady Mott have been able of late to indulge more frequently their affection for the game of croquet, albeit on a somewhat bumpy lawn.

The sheer variety and scope of Sir Nevill's contributions to modern physics defies summary description. The twentieth century has been a heroic age for physics; the Cavendish Laboratory at Cambridge will always, I believe, be remembered as one of those awesome places where man's knowledge of the universe was fundamentally altered. Sir Nevill Mott stands very firmly in a line of truly distinguished men of science who have staked out the claims of Great Britain to be regarded as one of the creative centres of research and discovery in our time. When Nevill Mott began his career in the 1920s, quantum mechanics was still in its infancy, but he very soon made his mark, with significant contributions to the theory of atomic scattering. One of his important achievements at this early period was the recognition of the role of symmetry in alpha-particle scattering by helium nuclei. Another distinguished scientist has written of Sir Nevill that 'perhaps nothing is more typical of his special quality than the way he has, from his beginnings as a mathematical physicist, moved speedily away from artificial models and towards an understanding of real substances and processes.' Sir Nevill's numerous achievements in solid state physics since the 1930s have shown this very clearly. I could at considerable length, expatiate on the topic of his work on the properties of metals and alloys—which yielded the enormously influential text known to physicists as Mott and Jones, only one of several influential texts which Sir Nevill has jointly written with other scholars. His research on insulators, his study of dislocations, and, from the 1950s onwards, his interest in liquid metals, impurity bands in semiconductors, and the glassy semiconductors—all these things, and many more, have left ineffaceable traces on the scientific history of our time. The common parlance of solid state physicists now includes such expressive terms as 'the Mott transition, Mott insulators, and Mott minimum metallic conductivity.' He is indeed a scientist of acute insight, an indefatigable experimenter who has attracted many distinguished experimenters to his laboratory.

Those who have collaborated over the years with Sir Nevill, and who have fallen beneath his spell—and they include members of our own Physics Department—testify time and time again, and with great warmth, to his immense charm and friendliness, his utter willingness to listen and to help, his openmindedness and receptivity to new ideas. As with so many truly great men of science, stories of his absentmindedness abound. If they are true, he is in no position to confirm or deny them. A research worker on the oxidation of materials, in a fit of excitement, is said to have told him: 'Sir Nevill, these results support your theory'—to which he is said to have replied: 'Just remind me of what my theory was!' Colleagues and friends testify, too, to a nice turn of humour. He has always been ready to see visitors, especially young people. A group of students visited him one day for a half-hour discussion which somehow turned into a session lasting all day. As they left, Sir Nevill said: 'Do come and see me again when you have a little *less* time to spend.' He is also believed to have described King Henry VIII as 'the greatest experimentalist of them all'.†

It would be entirely wrong, of course, to present a picture of someone so wholly wrapped up in his researches that he had no time for the wider world. Sir Nevill has played a full part outside the laboratory, as well: President of the International Union of Physics in the 1950s; a member of the Central Advisory Council for Education in England; a member of the academic planning committee for what presently became the University of Sussex; and more recently he has taken part in the work of the Physics Education Committee of the Royal Society and Institute of Physics. These

---

†At a conference dinner in the hall of Trinity College, standing below Holbein's portrait of the monarch.

are a mere handful of the public services which Sir Nevill Mott has made to the wider cause of science and to scientific education.

I find it hard, my Lord, to sum up so manifold and varied an achievement. The University delights to add its own small meed of praise by honouring this outstanding physicist, the ornament of his profession, his country, and the science of his century.

My Lord and Chancellor, I present to you Nevill Francis Mott

CHAPTER NINETEEN

Taylor & Francis

In Chapter 11 I have described how, after the war, I became interested in scientific publishing and was editor of the *Philosophical Magazine*, a scientific journal published by Taylor & Francis† since 1798. 'Philosophical' of course meant Natural Philosophy, what we now call science. The managing director when I first knew the firm was George Courtney-Coffey, an engaging character with considerable charisma and shrewdness but with little idea of what science was about. He depended on his advisors, such as myself and one or two people whom I introduced to his Board of Directors. We started some new journals. When I moved to Cambridge he kept on pressing me to join his Board, but I saw no reason to do that; I had plenty on my plate and enough income. However, once a year or so he and I entertained authors, editors and senior members of his staff to sumptuous 'Phil Mag lunches' in my college, where the best that our excellent kitchen could provide was consumed. In this and other ways we made friends and ensured that the company was well known in the scientific world.

However, when I saw that retirement was at hand, I changed my mind and joined. Courtney-Coffey was both chairman and managing director at the time, but sadly after a stroke his powers were rapidly waning and my first—and unpleasant—task was to persuade him to resign. After that I soon found myself chairman, a post which I held till my 70th birthday in 1975. Since then I have been President of the Company, with the right to attend board meetings but without a vote.

Of course after retirement the money was welcome, but apart from that there have been few experiences that I have enjoyed more. I found the atmosphere friendly; the board, containing both academics and full-time members, worked together excellently. I felt the same satisfaction in working with people from a totally different background as I did when working with the professional soldiers in my time at AA command. During a lengthy postal strike board members used to drive a van

† For a history of Taylor & Francis see ref. 18.

over to Holland to get our journals into the post for delivery abroad. Possibly to assign this job to anyone below board level would have been strike breaking. During the three-day week resulting from the miners' strike during Mr. Heath's government we kept going, printing the books and journals in the three days and examination papers, for which we had special permission, in the other two. A major decision which was before the board during my chairmanship was whether to abandon printing, concentrating on publishing and sending our work to the cheapest printer, or whether to buy a new factory and try to build up an economic production plant. We did the latter, our factory being in Basingstoke. Through many ups and downs we have remained profitable.

The firm exports 80% of its product. When it came to fixing the dividend, it was always a question of how little we could decently pay and how much we could keep for expanding our business, new plant and starting new journals. We employ about 200 people. I must say that a firm like ours, with little encouragement either from government or the unions, seems to me part of the 'acceptable side of capitalism', and I cannot imagine it operating as part of a nationalized publishing industry. Applied to a firm like ours, the idea that the sole aim of industry is to make profits is nonsense. Our aim is to survive, to grow and do a job that we can be proud of.

I should perhaps say one more thing about the *Philosophical Magazine*; I have myself published most of my papers on disordered materials there, and this means they were not necessarily refereed. If I had sent them to another journal and submitted to refereeing, doubtless many of them would have been better papers, but I am sincerely doubtful whether some of them would have been published at all! My ideas on a mobility edge, a minimum metallic conductivity and so on were hardly accepted by anyone at first. All I could say was that this is what a simple theory predicted, even if it was against most people's intuition based on their experience of other problems. It was some years before the experimental evidence for my ideas began to accumulate.

CHAPTER TWENTY

Talented Children

In 1965 the Royal Society set up an education committee to consider the teaching of science in schools. Since weakness there could affect the number and quality of the nation's future scientists, it seemed to the officers of the Society that school education was one of their proper concerns. I served on this committee from its beginning, retiring in 1982, and was chairman of several working parties on science in primary schools, science in colleges of education and the shortage of mathematics teachers. Finally I was chairman of a group for which I wrote a report in 1979 entitled *Science and the Organization of Schools in England; implications for the needs of talented children*. In this I was greatly helped by the Education Secretary of the Society, Donald Harlow; together we visited many schools in different parts of the country. The investigation was started because some members of the committee believed that talented would-be scientists would not get the attention and stimulus they needed in every comprehensive school. It was a time when mixed ability teaching was much favoured though certainly less frequently practised, and with priority recommended for the needs of the average child. At the same time there was controversy between those who believed that schools should teach O-levels (as they then were) in the three separate sciences, physics, chemistry and biology, and those who favoured an integrated course. A report issued by the Association for Science Education, entitled *Alternatives for Science Education*, favoured a pattern which, in our view, was more about science as a social activity than a course to enable students to start a career in this subject.

My report was controversial. We appreciated the needs of the average child, but thought that mixed ability teaching up to the age of 16 could be successful only with exceptionally able teachers. Our working party believed that as far as possible talented children should be allowed to advance at their own pace, and with this in view favoured 'setting' in science and mathematics from 13 years onwards. With such a system a child could find himself in a high set for one subject and a lower one for another. It would be possible to arrange this only if the school was above a certain size, perhaps 1000 children, and for this reason we were critical of the small comprehensive schools. I was able to see some schools in Sweden, where the teachers I met were unhappy with the mixed ability teaching in science imposed there.

The Royal Society arranged a meeting with teachers to discuss the report, some of whom attacked it as elitist. However, in a later and fuller investiagtion of science teaching under Sir Harry Pitt, it was not without its influence.

My experience with this report led me to organize, in co-operation with a Trust (the Leonardo Trust, devoted to the needs of talented children) a conference at Caius College in September 1981 on *Gifted children and their contemporaries*. Two members of parliament, a Vice-Chancellor, Mr. John Healy of the Association of Science Education, Mr. Brian Jackson and many other took part. My opening address, influenced by some adverse criticism of my Royal Society report, is reproduced in Appendix 2.

CHAPTER TWENTY ONE
Anti-Concorde

In 1967 a Cambridge friend (W. H. Thorpe FRS, professor of animal ethology) brought to my notice a leter in the Times announcing the existence of an 'anti-Concorde project', a group of some hundreds of people who wished to oppose the development of the supersonic aircraft Concorde. These people felt that it was an abuse of technology. The costs were escalating far above the original estimate; the machine would be extremely extravagant in fuel; and above all an aircraft flying above the speed of sound must necessarily inflict a sonic boom on the people over whom it flies. Spokesmen from the aircraft industry were confident that the public would learn to live with this, but those who backed the project felt that this was an intolerable and unnecessary infliction, in the interests of a few business men in a hurry. In fact supersonic flight has been limited to flight-paths over the sea. Financially Concorde has been a disaster and none have been sold outside the countries of origin, Britain and France. As is well known, the Wilson government made several attempts to cancel it, but could not get the agreement of our French partners.

The organizer of the project was Mr. Richard Wiggs, a former teacher of handicapped children. He and his group published full-page advertisements in the press, putting the case against Concorde, and paying for them through contributions from those who read the advertisement and agreed with it. He did not have much difficulty in raising the money, but there was little left over for a salary for himself.

In July 1967 I saw the following letter in the Times:

Anti-Concord project
From Mr. Richard Wiggs
Sir, Miss Pamela Hansford Johnson and Sir Alec Guinness (July 5 and 10), and many other readers of 'The Times' may be glad to know of the existence of the Anti-Concord Project, which has been founded by a group of some hundreds of people including scientists, artists, business men, civil servants, farmers, housewives,

professors, M.P.s etc. who are concerned and alarmed at the efforts being made to develop supersonic aircraft.

We see this as a clear case of a choice having to be made—is technology to be sanely controlled or is it to be allowed increasingly to degrade and destroy our environment?

Our immediate aims are to help to create in Britain a climate of public opinion in which it will be possible for the Government to terminate work upon the Concord, and to press the Government to make this decision. Our further aim (in co-operation with similar movements in other countries) is to help to bring about the banning of supersonic transports internationally.

We shall be glad to hear from people who agree with these aims.

Yours faithfully,
RICHARD WIGGS, Convener

and wrote to him as follows:

15.7.67

Dear Mr. Wiggs,
With reference to your letter in the Times, may I say that I agree with your aims and would be glad to help.

I then became chairman of the group and we met regularly in my Cambridge house. We had allies in the USA who successfully opposed the development of supersonic transport there.

Concorde is still flying to New York and Washington, supersonic only over the ocean. But no more are being made, and it looks as if eventually the aircraft must be phased out. Certainly supersonic aircraft have little future, and are now unlikely to become the scourge that we feared.

CHAPTER TWENTY TWO
Religion

In an earlier chapter I have said that I was not brought up in any religion, and at home with my parents we did not go to church. At Clifton College, however, Sunday chapel was compulsory and I left school very familiar with the prayer book service, and with an enduring affection for it, which was also strongly felt by Ruth. None the less, until about my fiftieth year, I do not remember having any interest in organized religion. However my return to Cambridge in 1954 to the Cavendish chair coincided with the appointment of Canon Mervyn Stockwood as vicar of Great Saint Mary's, the university church. Formerly Stockwood had been in charge of a slum parish in Bristol and I knew him fairly well; he had in fact officiated there at the marriage service of my sister-in-law to Jacques Friedel, the French physicist who was working with me. In Cambridge I invited him to talk to a class in the Cavendish where, once a week, we exposed our physics students to subjects outside science. Services in Great St. Mary's had attracted very small congregations before he came, but he soon changed that. I remember attending the church from time to time to see what he was up to, and hearing a very effective sermon about the Suez crisis. He left to be Bishop of Southwark, and his autobiography[19] records a letter from me saying how much I would regret it if he left. This letter is reproduced below. Reading it now I am reminded of Sir Philip Morris's saying that 'Cambridge is a cold, dead place'—and my belief that he said this because there was no-one at the top to express what it stood for. I felt that Mervyn and his church could do that. It was perhaps a pity that he left and became a bishop, and grew so disillusioned with institutional Christianity.

> I consulted friends; most advised me to stay. I had a letter from Professor Nevill Mott of the Cavendish which almost determined the issue:

> > I will try to put in writing some of my feelings about Great St Mary's, particularly as it affects senior

members of the university. You are a born teacher. You can put, or cause to be put, the pros and cons of Christianity, of Suez or of Wolfenden, to our young men, and to older men too, in a way which leads them to think the matter out for themselves. In this you are *very* much part of the university—a man who knows how to put ethical issues before us without hiding any aspect of their intellectual content. I think it important for those of us who teach in the university that there should be a focus for our aims and aspirations such as your church provides. We are divided in Cambridge into our inward-looking colleges, our ambitious departments and our various cliques. Apart from some devoted administrators and our ceremonial in the Senate House, what is there to remind us that we are a university with a purpose in the world? Your church does that—a perpetual reminder that values matter and that our job is to teach men to think about them. I know this is valuable to me and I am not the only one who thinks so.

Mott's letter, while too generous in his estimate of my work, caused me to write to Archbishop Fisher to tell him that I thought I should remain in Cambridge. He asked me to Lambeth to talk. 'Now Stockwood,' he said, 'sit down and tell me your reasons for not going to Southwark. There is no hurry; take as long as you like. Then I'll tell you why you're wrong.'
(*Chanctonbury Ring*—the autobiography of Mervyn Stockwood)

My letters to my parents record that Stockwood had suggested several times that I should preach or speak in his church. Eventually in 1957 he arranged for scientists to give a series of lectures there on Sunday evenings, and I agreed to give the first of these. In 1959 these lectures were published by the SCM press under the title *Religion and the Scientists*. Reading my lecture again after 28 years, I find several views expressed that I still hold. I must have done a lot of reading for this lecture—probably the first time that I had thought at all deeply about religion, and its relation to the kind of science in which I had made my career. In the lecture I emphasize the distinction between scientific and religious truth, and also, at the end, my belief that religion is based on the history of Christianity and that one should not change the words of the church services to bring it up to date. The lecture is perhaps repetitive, but I had to fill up an hour of speaking time. It is reproduced in Appendix 3.

However, in spite of my liking for the traditional service, I still did not often go to Church. But when I was Master of Caius, I normally attended the very beautiful Chapel service there at 6.15 on Sundays, sitting with Ruth in seats reserved for the Master and his spouse. I could not help regretting that so few of the Fellows came too, feeling that, whatever their beliefs, in the chapel they could at least join in reverence for the traditions that had created our college. When I resigned the mastership, Ruth, who had become an Anglican before the war, went regularly to Little St. Mary's—a well-attended church of the Anglo-Catholic persuasion, and I usually went with her. Since moving to the village of Aspley Guise in 1980, I have attended church there fairly regularly, finding it helpful to worship God in company with other people. If some things that are repeated in the creed do not correspond with what I believe, such as 'born of the Virgin Mary', I accept what is said because to me the Christian religion is the sum of the beliefs of Christians through the ages, not only those of our present generation. For this reason, as for many others, I much prefer the service of the 1662 prayer book, rather than the modernized versions such as Rite A. The old versions, with their beautiful language, express to me the timelessness of religious belief.

When in 1977 Ruth and I visited the University of Tübingen (see Chapter 16) we heard several speeches, including one by Professor Hans Küng, and we were privileged to meet him afterwards at lunch. Professor Küng was kind enough to write and to send me some of his books. I have been much influenced by his *On being a Christian*—by the doctrine that one cannot prove anything, but that faith is a matter of trust, and by his treatment of the Resurrection in a way that does not involve the movement of molecules.

In a series of broadcasts by Gerald Priestland, I came across the doctrine—learned in his case from the Bishop Taylor of Winchester—that the God of Christianity may not be omnipotent, and is only able to work through men by love. This is very attractive to me. It leads me to think that one cannot pray for peace, or the end of a famine, or for health for someone else, with any expectation that such intercession in any prayers will be granted. One can pray only for wisdom to know what is a right action for oneself, and for strength to do it, and for insight towards God. Sometimes I cannot help praying for those other things, but without any real belief that such prayers can be effective, except by enhancing one's own concern. The problem of how a loving God can, if omnipotent, allow the cruelty that there is in this world is of course an old one. For the problems like the holocaust in Germany,

Hiroshima, famines, handicapped children, I can find no other solution but to deny omnipotence. I was struck by the view of omnipotence which I found in a book by Karl Popper[20] which I reviewed as follows:

This book is one of three volumes which comprise Sir Karl Popper's long awaited *Postscript*. The others are *Realism and the Aim of Science*, (to appear soon) and *Quantum Theory and Schism in Physics* which is reviewed also in this number. It was written more than twenty five years ago and never before published, though—say the publishers—it has circulated in manuscript and galley proof and made a deep impact on contemporary philosophy and science. It is edited by W.W. Bartley, Professor of Philosophy in California State University.

The present reviewer, like most other practising scientists, has been deeply influenced by some of Popper's doctrines, particularly that which asserts that scientific theories can be falsified but not proved. But the case made in this book comes as a surprise. To most of us who throughout our careers have used quantum mechanics to describe and predict phenomena in physics and chemistry, quantum mechanics implies the uncertainty principle and thus indeterminism; if the uncertainty principle broke down, quantum mechanics would be falsified. Classical Newtonian mechanics, on the other hand, is usually considered as deterministic; in it the future is predictable from the past. In some way, not entirely clear, quantum mechanics has allowed us to escape from this mechanistic view of nature.

This is not what Popper maintains. 'I personally believe that the doctrine of indeterminism is true, and that determinism is completely baseless', he writes (p.41). This is no surprise. But what is a surprise is the cogent arguments that he gives for the belief that indeterminism is valid for Newtonian mechanics too—using the kind of argument made familiar in the discussion of Heisenberg's uncertainty principle. He argues that determinism means that, if we know the initial condition sufficiently exactly, we can predict the future, and that if we need to know the future with a given degree of accuracy, we must be able to say how accurately we must know the present. Even in Newtonian mechanics, since for the interaction of a large number of bodies, we have no analytical solution of the equations, he argues that this is not so and that as systems get more and more complicated so does determinism get less and less a good approximation of the truth. Cause and effect survive his analysis; if his cat jumps on the breakfast table in the morning, the cause may be that it likes the smell of milk, and he can predict it because it does it every morning. But determinism would demand that it should be possible to predict exactly on what part of the table the cat will land; it is this that Popper denies. There is of course much more in the argument which he develops. A section on 'the impossibility of Self-Prediction' shows that, if you set out to investigate yourself, you change yourself, just as it is with Heisenberg's uncertainty principle and the gamma-ray microscope.

This is all about what Popper calls scientific determinism. In a chapter on metaphysical determinism, he claims to have shown that complete prediction is impossible from within the world—but that there remains open the possibility that the world, with everything in it, is completely determined if seen *from without*—perhaps by the Deity. Is this, he asks, arguable? He says that it is. It is perhaps linked with the strong intuitive belief that many of us have, and have had through recent history, that Newtonian mechanics is deterministic and that—in spite of Popper—only quantum mechanics shows us how to escape. But Popper does not say this. He says that both metaphysical determinism and metaphysical indeterminism are irrefutable, and so he must then argue the case for the latter on grounds of common experience, common sense and a long conversation with Einstein which failed to shake him in his conviction that God does not play dice. This perhaps is not physics and not appropriate for review by a physicist in *Contemporary Physics*. However, if you are convinced by Popper and, or, by quantum mechanics that scientific determinism is not valid, and if as seems reasonable you follow him against metaphysical determinism, then that does tell you something about the Deity, as Popper makes clear enough. In so much of our church services, particularly in the hymns, the Deity is described as almighty, omnipotent, omniscient. 'Luther and Calvin', Popper reminds us, 'bought determinism because it is connected with the ideas of divine omnipotence—complete power to determine the future—and of divine omniscience which implies that the future is known to God now, and therefore knowable in advance, and fixed in advance'.

If this is not so, what about omnipotence and omniscience? Perhaps the days when physics can impinge on theology are not over. But certainly theology will be more welcoming than was the case in the days of Darwin and Huxley. It is perhaps significant that in some popular accounts of Religion, for instance Gerald Priestland's *Priestland's Progress*, we hear of theologians who deny omnipotence and say that the only power of God is through His interaction, through love, with man and man's response.

On such matters Popper has little to say in this book. But he writes (in a footnote) 'If everything is known to God; then the future is known; it is therefore fixed in advance, and unalterable, even by God Himself. I shall not discuss here the ethical difficulties involved in the doctrine of divine omnipotence, such as the ethical problem of whether it is not evil to teach the adulation of power'.

I agree with Popper. I do not worship omnipotence. God is the principle of good, and we must seek him as we can. But what He has to do with the creator of the universe I do not understand; perhaps the laws of physics are such that they could not be otherwise than they are. And God reveals himself to us through men, not through the stars. We in the West accept that the supreme revelation is through Jesus, and what his followers have recorded.

In Aspley Guise, where I now live, I have taken a considerable part in the life of the local church. I have given several addresses, covering much the same subjects as my lecture (see Appendix 3), and on the contrast between Rite A, the modernized communion service in the Alternative Service Prayer Book, for me a *bête-noir*, and the Book of Common Prayer, on miracles and on the theology of Hans Küng. I have also organized a study group on the Christian attitude to nuclear weapons, seeking to persuade people that there is an alternative to unilateral disarmament. Our conclusions (April 1985) are set out below.

A suggested Christian approach to nuclear weapons

This is a redraft of a paper that was circulated to the St. Botolphs discussion group on March 13, and is prepared for consideration by other parishes. Though I would be far from claiming that it represents everyone's point of view, I think that few if any strongly dissented from it.

What do we mean by a Christian approach? Reading the Sermon on the Mount, one knows that here is a way of life—'love your enemies, do good to them that hate you'—that can be adopted today and in fact is adopted by certain persons, for instance in religious orders. It can lead to a serenity which is shiningly apparent to those who meet them. But I believe, following I think most theologians, that Jesus, God and man though he is held to be, was sufficiently man to have believed, in the spirit of his time, that the Parousia (the second coming) was imminent. There was therefore little need to think about government, law and order, politics and the men involved in such matters. In the recorded sayings of Jesus and of St. Paul, there seems little advice to the Roman soldier who had to defend the frontiers against the barbarians.

It seems to me, then, with regard to pacifism, there are and must be those who feel they can have no part in war of any kind. Such people, probably, do not enter the Civil Service where they may find themselves in the Ministry of Defence, or the electronics industry which produces parts of guided missiles. If they did, then their obedience to the command 'resist not evil' would imply that they would have to advocate such a policy not only for themselves but also for all other citizens, who might wish to be defended, as most of us did in the war against Hitler.

What then can we say to those who have responsibility of any kind in these matters, or who seek to form opinion about them?

I will quote from an article by Michael Howard on the dilemma facing Christian statesmen, men who have to go on doing their job and who may be members of the church as well:

'Do they incur excommunication by continuing to do what is, by any standards, necessary work? They can of course always escape from their dilemmas by resigning, admitting that a certain action may be necessary in the context of the power situation in which they find themselves, but being unable to square it with their ethical principles, and regarding the latter as overriding. Few do. The cynic will attribute this to simple lust for power, but the explanation is seldom as easy as that. The statesman knows that somebody has to take the decision, and to refuse to do so is an abdication of responsibilities deliberately assumed. Pontius Pilate is an unattractive figure for Christians, not because he did his duty and firmly took a disagreeable decision, but because he failed to do so, taking water and washing his hands saying: 'I am innocent of the blood of this just person; see ye to it.' President Truman, staunchly accepting responsiblity for decisions of unimaginable consequence, is likely to occupy a more comfortable place even of the Christian Purgatory.'

We may not envy the position of the statesman and we may not always agree with decisions made; but we cannot, if we are seeking to exercise our moral responsibility, condemn him for being willing to accept it.

So—for most of us—the Sermon on the Mount is something in the background, and we can thank God that there are people who accept it, and revere them as the salt of the earth. But we know that the world needs the others, the businessmen, the politicians, the policy-maker. They need the people who can—and must—take decisions on national defence—and *any* decision short of absolute pacifism is not compatible with the Sermon on the Mount. Also, if they were pacifists, they would not be in the job they have. What can we wish for them? Can we do better than ask them to use their intellect—a God-given faculty—to order things in such a way that a nuclear war is least likely to happen?

If then we do not take the pacifist point of view, I do not believe that there is a specifically Christian answer to the problems posed by nuclear weapons. We must expect disagreement. The following are simply my own views.

1. The potential Soviet threat is a real one. Few people believe that, if NATO got rid of the bomb, the Russians would immed-iately march in; on the other hand, if we were seen to be defenceless, western Europe would increasingly come under their influence.

2. None the less, everthing should be done to increase trust and to moderate the cold war.

3. The present policy of NATO is 'flexible response' which means retaining the option of first use of nuclear weapons in response to a Russian non-nuclear attack. Most authorities believe that, once this threshold is crossed, there is every likelihood of escalation to the ultimate catastrophe. I am forced, therefore, to feel that a 'no first use' posture is the moral one; but I am very well aware of the arguments against it, for instance that it might make conventional war more likely. I think one has to examine the pro-blem with an open mind, asking oneself what policy is the most likely to lead to a peaceful future.

4. I am attracted by the attitude of the organisation 'Just Defence', which advocates a credible non-nuclear unprovocative defence of western Europe, retaining for a time being a submarine-based 'minimum deterrent', the purpose of which would be left rather vague, but only to be used if the Russians used nuclear weapons against the West.

5. As regards the American alliance, I do not think one can do without it, however critical one may be of some aspects of American policy.

6. I see no reason for the acquisition of Trident, as long as we retain the American alliance.

7. Cruise is no more dangerous than the (American) weapons it replaces, and perhaps less so, because it can be dispersed.

8. The 'star wars' proposal (Strategic Defence Initiative) should be opposed. One suspects that defence of the civilian population is impossible, and that the real objective, by defending missile sites, is to achieve superiority over the Soviet Union. Its result will be a major escalation of the arms race; which, if nothing else, would be ruinously expensive.

CHAPTER TWENTY THREE

Looking Back

In 1980 Ruth and I left Cambridge, moving to a smaller house in the village of Aspley Guise near Woburn. We did this to be near my daughter Alice, her husband Michael Crampin and their three children; he is a professor of mathematics in the Open University. I retained a room in the Cavendish, spending one or two days a week there, continuing to work on the problems that have occupied me for the last twenty years and discussing them with colleagues. I was appointed Adrian Fellow at the University of Leicester, so that I might visit Ted Davis, the co-author of my major book, now professor of physics there. I travelled a lot, to Israel to a conference arranged by Moshe Kaveh, to the USA and to Rome for the conference on amorphous semiconductors in 1985.

Looking back on a career in physics, I realize how much I have been at the mercy of chance. There was never any doubt that I would be a physicist; it was obvious to me that this was the thing I could do best, and I never considered any other career. Because I started research in Cambridge, with Rutherford at the peak of his powers, and at the moment when quantum mechanics was first formulated, it was natural to apply these new ideas to Rutherford's nucleus. Four months with Bohr showed me what theoretical physics could be, a continual exchange of ideas, a social activity and not one to pursue in the solitude of a college room. I then knew what kind of department I wanted to run, and in Bristol in 1933 I got the opportunity. And there I turned to solid state physics, not so much by deliberate choice but because it was already going on there. My experience in the war made me want to build up a department that could help industry, and in this at Bristol from 1945 onwards I had some success, and perhaps also at Cambridge. For twelve years, after my retirement, I have been a consultant to the Plessey research laboratory at Caswell, and I have also been consultant to Harwell, The Royal Radar Establishment at Malvern and Thomson CSF in Paris.

As regards my work on non-crystalline semiconductors, to find this

Figure 20. The author, taken at the triannual meeting of Nobel prize winners in Physics in Lindau (FR Germany), July 1985.

subject in the late 1960s, in a state when sophisticated mathematics had not been applied to it, was a piece of luck. I got in first. From a scientific point of view, it was like 1933–39 over again; there was everything to explain with simple mathematics and simple models. From the 1970s onwards, the subject became fashionable; the theory became difficult and the experiments expensive.

The greatest pleasure in a life in research is putting others on the way to success. How some of them felt about it is shown in the letters that they wrote me on the occasion of my Nobel prize (see Chapter 16). Exchanging ideas is always a joy, whether the ideas are in politics, religion or physics, and a physicist has that privilege often. Then another satisfaction is making a theory that predicts a new phenomenon, and finding it confirmed by someone's experiment. Argument and disagreements with colleagues are usually pleasant—but occasionally they can become embittered. Any scientist, myself or another, can become so enamoured of his brain child that he resents criticism. I remember a paper given at a conference, part of which was on the theme that (in a restricted area) 'Mott is always wrong'. I did not like it, but agreed that I am not always right, and was in fact wrong about one of my ideas criticized there.

I have described in this book how I chose subjects for research. Essentially I look at what experiments my colleagues are doing, ask how I can help and whether I have the ability to do so. If I see a problem and a collaborator and I feel we could do something together, we have a shot at it. The collaborator can be in any country.

I am glad that I did not spend all my life in Cambridge, glad that I was pulled out of pure research for the war period and that experiences with the problems of school education and in business (Taylor & Francis) have come my way. My biggest job was, of course, to run the Cavendish Laboratory. Again I was fortunate in that it was expanding during my tenure of office, and that money was available. If in one year there was no position for a talented colleague, the next year there probably was. Now it is not like that, and harder choices have to be made.

During my time as a physicist, the world of physics has completely changed. Before the war, still more before 1933, the small community of atomic scientists all knew each other personally. Europe was the centre of physics, with Göttingen, Copenhagen, Cambridge and Paris outstanding. After Hitler came to power, the flood of emigrants from Europe greatly strengthened the subject here and in the USA. But we remained a small community, and what we were doing was a mystery to most people.

The war, the Manhattan Project and the bomb changed everything. Everyone knew only too well what physics could do. Physics split up into its different branches, each with its own journals and specialists and the number of physicists increased enormously. We did not all know each other any more. The cost of research in physics escalated; what anyone could do depended on access to funds. The computer became essential and greatly extended the possibilities.

At the time of writing this book money is short, and I look with anxiety on the future of the two laboratories over which I have presided. In 1982 a rather perceptive article about the Cavendish appeared in *The Economist*; I reproduce part of this (in Appendix 4), as it describes fairly enough what I tried to do in the Cavendish and how my successors saw it. I was appointed in 1954 to head the laboratory until I reached the retiring age; my successors were not, it being thought that a limited tenure would be enough. Now, from 1984, with the appointment of Sir Sam Edwards until his retirement, the University has gone back to the old pattern. In a difficult period, this may be the right decision.

The main job of a head of department, I would say, is to back talent and *not* to be fair between the different groups. Looking back, I dare to feel I made the right choices, either by skill or luck. I think it was a pretty healthy laboratory that I handed over to my successor in 1971. J. G. Crowther's book quotes me as calling it, 'one good laboratory among many'. I'm sure that this is what I felt—in the sense that I wanted to deny strongly that there *ought* to be anything special about Cavendish nuclear or any other kind of physics. But actually when I retired in 1971 it struck me as being well above any other university laboratory in the country.

It was also a happy laboratory, or so it seemed to me. I felt (like President Truman) that 'the buck stops here', and if anyone wanted to come and grumble to me, they could. Also I was enormously helped by the Academic Secretaries of the Department, Ian Nicol, and following him John Deakin, to whom my debt is very great. I am also very grateful to my two secretaries, Mary Brown and Shirley Fieldhouse, in Cambridge. And—I need hardly say—the 'Cavendish wives' club', organized by my wife, in which wives of staff, research students and particularly visitors met in the laboratory tea room and heard talks from eminent people, added greatly to the feeling of a happy community.

APPENDIX ONE

Science and Modern Languages

N. F. Mott

The following is the text of the Presidential Address delivered on 29th December, 1954, at the College of Preceptors, London, by the President of the Modern Languages Association for 1955, Professor N.F. Mott, M.A., D.Sc. (Honoris Causa) of the Universities of Louvain, Grenoble and Paris, F.R.S., Cavendish Professor of Experimental Physics in the University of Cambridge.

I am very honoured and very happy that you have chosen a scientist to be your president this year, because it gives me a chance to say something about the relation between the teaching of science and the teaching of modern languages. I believe that scientists ought to learn modern languages; I know that you, the teachers in our schools and universities, signally fail to teach modern languages to only too many of our scientists; and I believe that this failure is not your fault. I believe it is ours, the fault of the scientific staff of the universities. So in my address to-day I have three things to say: why I think that scientists should learn foreign languages at school and university; whose fault it is that they fail to do so, and what ought to be done about it.

Why do I think that scientists ought to learn foreign languages? Not by any means only because it is useful to be able to read scientific works in foreign languages. Of course it is useful. But it is by no means so useful as it was when I, for instance, began research in physics. When I was a student at Cambridge, that revolution in our ideas about nature which we call quantum mechanics was just beginning. It had its origin in Denmark and Germany, and the papers about it were written in German. If we wanted to read about it we learned to read German, and if we wanted to talk about it we learned to talk German. But now, though German science has taken its part in the remarkable recovery of Western Germany, I think it is fair to say that that country is no longer the centre of any main branch of science. Certainly neither German nor French is the scientists' language, the language of international conferences, for instance. This language is English, particularly English spoken after the American manner. I was much struck, in Mr. Robert Birley's presidential address to this association some years ago, with his plea for a Koine or common language to make European Union stronger, and also by his frank admission that this was bound to be English, because of the importance of the United States in the world to-day and the extent to which English is already taught in the schools of Europe. You could substitute for Europe the whole non-communist world. In all this world English is the common language for scientific intercourse. I have recently been at a scientific conference in Japan; nearly all the large scientific output of that country is printed in English. Indian science is published in English, so is most of that of Scandinavia and Holland, and much of that of Italy. There is a tendency even in France and Germany to publish books and summarizing articles in English, so as to make their work accessible to all western scientists. In fact, it is only the man who wants to probe into Russian science who really needs to know a foreign language; and if the Iron Curtain remains porous enough to allow, as at present, a few copies of their main scientific publications to trickle through, Russian may become in a decade or two the essential foreign language for the English scientist. And if, as at present, it allows a few of their scientists to come to our conferences and a few of ours to theirs, then I can hardly describe how a knowledge of their language would break down barriers and help us to speak of the place of science in our common world.

However, I am not going to make a plea for the teaching of Russian to all English schoolboys in the scientific Sixth Forms; as your president I should prefer to deal with the practicable rather than the ideal. I could add that the ambitious young scientist who needs to know Russian for his work will teach himself Russian, just as my generation taught itself German. And in like manner, if, as we all hope, there is a great renaissance of science in continental Europe, perhaps around the new European Centre for Nuclear Studies at Geneva, the young men who will flock to these centres will be able to pick up the necessary languages, whatever they have been taught at school. It will be easier for them if they have learned a language at school, as they will have learned how to learn languages. But none the less, it is not on grounds as narrowly utilitarian as these that I advocate the teaching of modern languages to scientists.

My reasons are rather broader. They have to do with the part which scientists are going to play in society. It is a truism that we live in an increasingly scientific age. In the next decades our life and indeed our survival is going to depend far more on science than it does at present. Let me give you only one example. This country is spending perhaps fifty million pounds each year on the development of atomic energy. Why are we doing this? There are many reasons, and doubtless some are military. It may be worth while for England to make its ten bombs to the Russians' hundred and the Americans' thousand—my numbers you will realize, have no validity; it may be thought worth while to have some technical skill in this field, some knowledge which we can trade with the Americans for some of theirs. But the other reason which justifies this great expenditure, the reason why I would support it, and why I hope you would, is that here we have the source of power of the future. Within the lifetime of our children our English coal will become scarce; in two hundred years it will be nearly gone; we need it for many things, for coke for smelting iron, for chemicals, for dyestuffs and for nylon. It is criminal just to burn it in furnaces and open grates at the rate we do. The sooner we can develop another source of power and save our coal the better for our children's future. Moreover, with the rising standard of life in countries hitherto primitive, there will be a growing demand for electric power from atomic energy. We hope that it will be England that can supply it. We hope that in the new industrial revolution it will be England that can export the equipment and the skill in this vital field, and that our ability to provide it will enable us to keep our place in the world. I think that the future of us all depends more on the back-room boys of atomic energy than you may think—on them, and on other scientists too.

How does this estimate of the future affect the job of the professional scientist? It is certain that there will be a lot of them who will not be back-room boys at all, but who will need to come out of their laboratories and exercise leadership in the outside world. And of course there will be a lot more who will—God bless them—always remain back-room boys. I remember during the war when so many scientists went from committee to committee, sat on the doorstep of a Minister and hobnobbed with generals, one always found in any project, on penetrating beyond the directors and the policy-makers, a chap called Bill in his shirt-sleeves getting on with the job. Neither science nor the community can get on without Bill. And whether Bill learns modern languages at school he will still be Bill, and I do not think it will make very much difference to his outlook on life.

However, we must have our policy-makers too. A community

which depends for its survival on science needs scientists who are articulate, who can tell their fellow men where science is going and what needs to be done about it. Such men are needed in industry, as managers and on boards of directors; they are needed in schools, as teachers and as headmasters; they are needed in the BBC and on the staffs of newspapers. They are needed above all in the government and among the government's advisers.

While I am speaking of scientists in the government and advisers to the government, I cannot resist the temptation to refer to the case of J. Robert Oppenheimer, now past history, you may feel, but none the less relevant to my theme. Oppenheimer has been and is one of the great Americans of his generation. He is a man brought up in the remote abstractions of theoretical physics, who, finding himself between the wars amid the unemployment and distress of the great depression, turned hopefully like so many of his contemporaries to Communism and then turned away with disgust at what he saw.. He is a man who then had the courage to accept the leadership of the Los Alamos laboratory where was built the first atomic bomb for America. He is a man who from the end of the war onwards devoted his powers to the attempt to secure the International Control of Atomic Energy, and to guide his country's atomic policy in the direction that would lead to security for America and to peace for mankind.

Well, as we all know, the community in which he lives has turned and rent him, raked over every detail of his private life, termed him a security risk and taken away his unrivalled power to serve his country and mankind. It is a story which future generations may well compare with the trial and death of Socrates.

Why did this happen? Is it because, unlike Oppenheimer, too many of us scientists are inarticulate, staying in our laboratories and doing there the work we love, and not coming out into the world and talking about it, and helping to form a sane public opinion about science and what it can do for mankind? I think that in part it is. I think that it is the duty of leaders in science to show that scientists are men who share the hopes and aspirations of common humanity, and who are desperately anxious that their work should be used to bring mankind through the jungle of the twentieth century to the sunlit uplands of the Churchillian phrase. Here then is another reason why we need educated scientists, scientific leaders, and a great many more of them than we now have.

If then we need scientists who are able to see beyond science and to help find the place of science in the community, what kind of education do we want them to have? I have indicated that I want for them a humaner education, an understanding of the hopes, the feelings and the behaviour of other men which a study of the humanities can give. I would like many of them to acquire this through the study of a modern language. Coming down now to questions of curriculum, I would like them to spend from a quarter to a third of their working hours studying a modern language, right up to the end of their time in the sixth form. I will now try to give you my reasons for believing that it is in the study of a modern language that most scientists will find what they need.

First of all I agree with what Mr. Birley said in his presidential address some years ago. I believe in Europe and I want England to give to it the leadership that Europe needs. I want those of our scientists who will be leaders to know Europe and to love Europe. It may be true—it probably is true—that English is becoming the lingua franca of Europe and of most of the non-Communist world. But the way to bring England and Europe together is certainly not to wait for this to happen. Let our young scientists be brought up with a knowledge of Europe, an interest in Europe, in how Europeans live and how they speak.

But my desire that our scientists should learn modern languages is by no means only because I hope that they will play their part in the development of European science. It is because I believe that a sound training in a modern language, continuing until the boy leaves school and occupying four or five periods a week, would provide the best possible training in those fields of experience outside his special discipline, designed for the budding scientists from among whom our scientific leaders will come. Let me first describe the narrow training that only too many of them have at present, and why it has come to be so narrow; and then give some further reasons why I would choose to add modern languages rather than any other course of study.

As regards the degree of specialization in our sixth forms, I have no quantitative information, and I can only give you an impression gained from a wide experience of interviewing candidates for admission to the physics department of the University of Bristol, when I was head of that department. We always asked a candidate what he did in addition to physics, mathematics and perhaps chemistry. Usually he said that he had been doing one period each week of English, one of civics, one of gym., or something equivalent. Very rarely did a candidate say that he had kept up the study of any foreign langauage, so rarely as to startle the interviewing committee and lead them to consider him with some special favour. I believe that boys who specialize in mathematics have an even narrower curriculum, mathematicians in schools and universities frequently believing, in common with classicists, that the study of their particular subject provides a complete education.

It is not immediately obvious why so many schools give their young scientists so narrow a training. There is certainly a lot to be said for letting a talented boy do what he wants to do, and a boy with a talent for science usually wants to do science. But very many of us in the schools and universities believe that the process has gone much too far. Perhaps not too far for the man of outstanding talent, the future Einstein or Rutherford or Oppenheimer. Such men can throw themselves into science and devote years of their youth to it and yet have energy and vitality left to absorb what they need of the rest of human experience. But it has gone much too far for the thousands of fine talented men and tens of thousands of honest ordinary men who are subjected to it. It is often said that the standard demanded by the universities is so high, the competition so keen that the schools simply have to sacrifice a broad education in the attempt to get their pupils in. But this is simply not true. The competition now is not keen. There are universities in Great Britain with more places than qualified applicants. In Bristol the reputation of the department and the university are good; the university there in most departments has four times as many applicants as places. In many colleges at Cambridge the ratio is even higher. But this means only that it is difficult to get into some departments of some universities, not that it is difficult to obtain a university education as such. In fact, in Bristol the department of physics sent out a questionnaire to discover what happened to all those students who were not accepted. The replies showed that ninety per cent. of them or more had secured admission somewhere else. In fact, university education is available to all who can reach the advanced level in two or three scientific subjects, not a really exacting test and certainly not one that demands concentration on science to the exclusion of all else. Moreover, many universities run an intermediate course which prepares men for the honours courses, and to which men and women with a more general education can be admitted.

Although it is not unduly difficult for a science student to obtain a university education, it is of course much harder to obtain an entrance scholarship, particularly an entrance scholarship to Cambridge or to Oxford. These are competitive, and however reasonable the syllabus, the standard is set by the quality and training of the competitors. With state scholarships and awards by local authorities available, the winner of an open scholarship does not have much more financial help than other students; but none the less the prestige of these awards remains very great. I believe that all too many science masters—and perhaps teachers of other subjects too—judge their success in their profession by the number of these scholarships which their pupils obtain; and that it is this which leads them to plan a curriculum in which little is done but science. And it should be emphasized that this curriculum affects not only the scholarship winners but the very much larger number of scientists who are educated in the same way along with the scholarship winners but who ultimately obtain a university education by other means. These are probably the people who are most harmed by it. The cure is in the hands of the college authorities and the major schools, working jointly; many of us are aware of the problem and one may hope that something may be done.

Let us then imagine ourselves in the world of a few years ahead, in which by the efforts of all of us, it has become the recognized thing that the sixth-form scientist should spend a third of his time on some subject other than science. What then should he do with this time? Well, we are, thank God, a free country and each school will make its own choice. But whatever else we do, don't let us put in a lot of snippets. Don't let us have two hours' scripture, two hours' civics, two hours' gym. Personally I would choose a language. And certainly I would choose a modern language. For the young scientist Latin or Greek simply will not do. A thorough classical education is a very fine thing and teachers of classics are often fine men and devoted teachers. But in my view the study of the classics loses most of its value unless you can take it to the point where the literature can be read with pleasure, and with the time available this is too much to hope for. Classics at a lower level, grammar, prose and translation, will teach the student accuracy and logic and train his memory; but he can learn accuracy and logic in his mathematics and any chemist who has studied the properties of the ninety-two elements and their compounds must have had his memory well developed. But the final objection to classics is that no young scientist will ever see the point of it; he will feel that it is just a thing which prevents him getting on with science, a vested interest of the classics master.

He will see the point of a modern language easily enough. He will know that he has to read the work of foreign scientists; he will want to travel abroad; as often as not he is something of an idealist, who believes in international co-operation, in which he believes that science has a part to play. The idea of learning a modern language, French or German or Russian especially, will make sense to him. This applies particularly to the articulate students, those born to have broad interests and to become leaders in science and in the community. But perhaps after all it applies also to the future back-room boy, the man whom I called Bill. Bill will have to express the results of his researches in English and how much better he will do this if he has learned a foreign language. Bill and his family will cross the Channel, and how much more fun they will have if they can order their dinner in French. Anyhow, can you tell at school who will be the Bills and who the Oppenheimers?

Suppose then we agree to teach our scientist in the sixth form a foreign language; what kind of instruction should we give him? Well, certainly not an education that includes literary criticism. Put good French or German before him, but don't teach him what he ought to think of it. Appreciation will come to him later if he has it in him to appreciate, but don't try to teach it to him while he is at school. Teach him to talk the language a little, to have confidence and to speak it loudly and clearly, and not to mumble. Tell him the why and wherefore of the language, explain the grammar to him. If you don't, he will feel it arbitrary, unscientific and not for him. Science students, and I believe many other boys, like to know how things work, and grammar shows them how the language works. Couple your teaching of the language, also, with some talk about the history and life of the country concerned. Need the barrier between the modern language and history teaching prevent this? Could you not go so far as to co-operate with the teachers of science and tell the boys and girls of the science sixth, in French, something of the life of a French scientist? Can you not tell them the story of the Curies, of Pasteur, of Henri Poincaré? Let them learn about Lavoisier, what he discovered and why he was guillotined, and why, at that period as in this, a great republic had no need of one of its greatest scientists.

May I also make a plea for Russian, in one or two schools at any rate? If even a few scientists could come up to the universities knowing the elements of Russian, I believe this would be a first step of real importance towards our understanding of the work and thought of that country.

Finally, ladies and gentlemen, I would like to express my belief in the very great importance of your work, especially as it concerns the boys and girls studying science. I hope that between us we can educate the men and women who will understand and grapple with the problems of the coming years.

In 1985 it came to my notice that the University of Sussex has a very popular course which combines science with European studies, the fourth year being spent abroad.

APPENDIX TWO

Gifted Children And Their Contemporaries

Professor Sir Nevill Mott, FRS

Introduction to a Conference held in Cambridge in September 1981.

To explain the title of this meeting, may I propose a subtitle: 'How do we enable exceptionally talented children to develop their full potential without detriment to the interests of the majority?' I believe that we should discuss this issue within the context of the state system of comprehensive schools of this country, though it goes without saying that the experience of other schools, especially those which deal with a wide range of abilities, will be relevant to our discussions.

Of course, the question we are going to discuss is not new. Any answer may depend on political and moral judgements of the resources that ought to be devoted to future specialists, and this issue has been widely discussed in the context of science and mathematics. Perhaps what is new in our meeting is an attempt to compare the science and maths problem with that in the teaching of modern languages, and we are very happy to see here several experts who will speak on this subject.

Although political and moral judgements may affect our judge-ment, I very much hope that our objective will be to see how far we can agree on these issues, whatever our political loyalties. We will all agree that talented children need, both for their own sakes and for that of the community, the widest opportunities to develop their talents. Here I would commend to you that paper submitted by the Head of Carisbrooke School, who describes movingly a school where *all* children are given the maximum opportunity. Mr. Cornall denies that the needs of society should play any role in designing the school curriculum, except in so far as they reflect the needs of individuals; I think this is fair enough, but society does depend in many ways on the more talented members; almost any paper from DES and the Inspectorate refers to the need to relate education to working life, at all levels of achievement. In this sense—I hope Mr. Cornall would agree—looking after the individual needs of the talented is in the interests of society. What is of course essential is that any attempt by teachers to identify the talented (as in the Oxford project), and give them special attention—should not miss out on children from homes where there is little background in literacy or numeracy, but very much the reverse.

I have come to the problem from the side of science and mathematics. A few years ago the Royal Society set up a working party to look into the needs of talented children so far as these subjects were concerned; by 'talented' we meant those likely to go on to any form of further education, and our discussion was about the Comprehensive Schools of the country. Our report has been distributed to this meeting. We came to the conclusions that, provided the school was big enough to ensure, after a diagnostic year or two (11–12 or 11–13), three or more 'sets', up to 16, the education offered could be very satisfactory. The report also discussed the relative advantages of retaining the three O-levels in physics, chemistry and biology, recognising that only a minority of students would take all three, against an integrated science course equivalent to two O-levels which might be compulsory for the whole school. The Royal Society Committee concerned with the future engineer, scientist and technician, had nothing to say about the science for the arts man, science for the citizen, science for the less talented. Neither has the Finniston Report 'Engineering our Future'. But that these are issues in the daily life of every science teacher is abundantly clear, as shown for instance by the report of the Association for Science Education, 'Alternatives for Science Education'. Whatever the merits of the proposals made in this report for the majority, and they may be great, very little is said about the future specialist, on whom according to Finniston, our industrial future will largely depend.

The Leonardo Trust concerns itself with the talented child, but I believe that this meeting should concern itself with all children. We certainly need good engineers, broad and humane in their formation; we also need an educated population which will understand something about the pros and cons of nuclear energy, pollution and the rest of it. Do we really know the best way of getting either? How, too, should the scarce resources in physics and mathematics teachers be divided between the two objectives—unless of course we believe there is no conflict of interests? If any good ideas come to the surface at this meeting, the Leonardo Trust is here to follow them up, arrange further discussions and publicity. One single weekend isn't going to change the world; I see this as the initiation and clarification of a debate, which will continue.

Of course, in the aspect of 'science for all' versus 'science for the future practitioner' our debate goes back at least to the beginning of the century. I recommend the book by E. W. Jenkins 'From Armstrong to Nuffield' (1979) to make this clear. I hope at this meeting we can have one or more views of what 'science for all' is, what it ought to be and how it differs (if it does) from what the future practitioner will need. On this I have heard many answers. Science is not a collection of facts, but a way of answering questions, and the 'answer' continually changes. Can schools get this across? Or is the main objective a capacity to think quantitatively? Science can be, for the child who comes from some backgrounds, a very necessary exercise in self-expression in English. What other objectives do we recognise, and are they different from those of the future engineer or scientist?

Mixed ability teaching is rare in this country after the first two years of secondary schooling. Do we have views on its desirability and if so in what subjects? Do we believe that, with the average teacher at any rate, it will greatly help the less gifted, or not hold back the more talented?

I turn now to modern languages. Here too, for a century or so there has been a running fight between those whose approach to language teaching is grounded in the great tradition of the classical languages and those who see modern foreign languages mainly in terms of preparing children for face-to-face communication. As the classical tradition has declined, the latter tendency has steadily gained the upper hand, rather more rapidly in recent years. At the same time the study of a modern foreign language has been put through the mill of comprehensivisation. From being a subject mainly confined to grammar schools, French, at any rate, is now offered to all, and will be treated as part of the 'core' curriculum in the early years of secondary education. As a result the 1970s have seen a 'crisis in modern languages'. Teachers were faced with failure and drop out on an alarming scale and have been giving most if not all of their thinking as to how to give a worthwhile experience of language learning to their new constituency. Since that in turn means concentration on the concrete and patently useful, there has been a con-siderable confusion of the two issues: culture versus communication and high ability versus low ability, which has been especially unhelpful in discussions of languages at sixth form level. The failure of modern language 'A' level entries to keep pace with the overall expansion of sixth forms has been a cause of concern. It does not seem that we have a framework of studies and options which appeal to bright young peo-ple, despite their increasing opportunities for travel and international professional contacts. It is time that we studied their needs, motivations and learning styles much more carefully, and thought about the balance of resources. (See Modern Languages in the Curriculum, Eric Hawkins, Cambridge 1981).

We shall hear of some experiments in the 11/14 age group to make Modern Languages meaningful. I would like to add that I was disappointed to find no word about modern languages in the Finniston Report.† It is of course unrealistic to expect schools to ensure that some future engineers have enough Portuguese to sell their products in Brazil. But schools could ensure that a high proportion of those going into such professions have gained confidence in at least one foreign language and know that they can learn another if they have to. How can we ensure this, how can we obtain the teachers and, once again, what are our objectives for the majority? Some ability to communicate, some comprehension of the grammar, or what?

† On education in engineering.

Adddress in Great St Mary's Church, Cambridge, 1957

This is a series of talks by men and women who teach science in Cambridge. That we have been asked to give these talks is a consequence of an emphasis on the intellectual content of Christianity and of a welcome to free inquiry. Canon Stockwood has not asked any of us whether we propose to speak for Christian dogma or against it. All that binds us together is the belief that religion is worthy of thought and not, as some of the more extreme humanists may assert, something that ought to be ignored in the affairs of this century. As long as we agree to this, and we must since we have come here to talk, we are free to say what we like.

We come then to Great St Mary's, a group of scientists free to speak in any way we like about the purpose for which this church exists. In doing this we are scientists speaking about religion. I would like to say at once that as scientists we cannot speak about religion with the authority of science or with any authority whatever. This is essential to what I have to say and, together with what I believe to be the separateness of religion and science, is my main theme. I would like to contrast my position in this pulpit with that of Canon Stockwood himself when he spoke last year in the Cavendish Laboratory on the Impact of Science on Religion. Canon Stockwood, though he was careful to describe what he had to say as an expression of opinion which would commit nobody but himself, none the less spoke with the authority of protestant religion when he showed us how science had impinged on Christian doctrine, purged it of beliefs which owed nothing to its founder and led to a sharpening and purifying of its theology. I believe that most of what he had to say would commend itself to his fellow churchmen.

My fellow speakers and I, on the other hand, appearing as scientists in a Christian church, can speak with no authority at all. For myself I can only speak of religious matters as they appear to one man, whose profession is science and whose outlook is doubtless much coloured by his profession. I have no guarantee that what I say will commend itself to any of my fellow scientists. When a scientist speaks on scientific matters, the position is quite different. On the significance of a recent discovery, a new meson or a radio wave from outer space, we may disagree; but we disagree secure in the knowledge that in a few years experiment and theory will clear the problem up, and on this matter we shall know the truth and agree about it. Those of you who are scientists and most of you who are not are well enough aware of the kind of truth to which I refer; let me call it scientific truth, based on experiment. Those of you who are not scientists may feel that it is an inferior kind of truth, based on experiments of limited accuracy and theories which may be overthrown, and lacking any quality that can be described as absolute. I do not agree with anyone who thinks this; the test of scientific truth, apart from its simplicity and elegance, is the success with which predictions can be made from it and technology built on it, and the wide consent which it subsequently commands. And as for scientific theories, the good ones are rarely overthrown; as science progresses and more accurate experiments are made the worst that can happen to a theory is that it is shown to be part of a wider pattern of things. This is what happened to Newton's laws when Einstein took them in hand. Nothing could be less accurate than J. C. Squire's addition to Pope's couplet,

> Nature and Nature's laws lay hid in night
> God said 'Let Newton be' and all was light,

which is

> It could not last; the Devil howling 'Ho
> Let Einstein be' restored the status quo.

Newton's laws, even if comprehended in the wider synthesis of Einstein, retain their beauty, their truth and their practical value, and all scientists are in agreement about this.

I need hardly emphasise that religious thought does not advance in this way through experiment and widening agreement, and one may hope that it never will. The religious field of thought attracts — I speak personally — by its essential uncertainties, its fields in which there is no agreement, its glimpses of those problems in which despite their intellectual fascination no certainty is possible, except, if you choose, through an act of faith. By these I mean the most fundamental problems, the existence of God, the nature of God and his relation to man. In this essential uncertainty religion contrasts with science, which deals with a more limited field where a kind of certainty can be obtained. A wise man in the twentieth century will train his mind so that he can walk in both these fields of human activity and aspiration. But though I think that one ought to be able to walk in both fields, I doubt whether one can say that they are both part of the same activity, or that it is easy to find an outlook on life that comprehends both. There seem to me to be few points where they touch and not much that is essential that either has contributed to the other. This perhaps is a controversial statement. Whitehead and many others have maintained that the Christian religion contributed to European thought that belief in natural law which enabled science to arise, and that an intellectual climate in which the Deity was an arbitrary despot would have been much less favourable. Much has been written on this theme. Canon Smethurst in his recent book *Modern Science and Christian Beliefs*[1] has a lot to say about it, and there one finds a rather surprising chapter heading, 'The necessity for science of belief in one rational God'. One can hardly doubt that Christianity which contributed so much to European culture helped the men who first turned their minds to science, as they certainly believed it did. But one must allow too some partial truth to the contrary Marxist hypothesis, that, as astonomy was stimulated by the needs of navigation, so science arose when it did in response to the economic needs of the time. In any case, once the movement was under way, by the beginning of the nineteenth century, I would say that religious thought has contributed nothing to the progress or method of science. Certainly non-Christian countries have had no difficulty in taking up and adding to science without finding it necessary to believe in one rational God. It is of course true that the scientist makes certain assumptions which may be felt to imply an act of faith akin to the belief in a rational God, such as that nature is regular and not arbitrary, that disciplined imagination controlled by experiment will tell us the truth about things, and that the advance of knowledge is worth while. Certainly too a scientist makes certain assumptions about the ethical behaviour of his fellow scientists, that they are honest and painstaking, that their results can be believed. Without this science cannot advance. But I do not believe that for most of us at this moment of history these are religious or even ethical judgments; they are judgments forced on us by the success of scientific thought and practice and necessary for anyone whose talents or ambitions lead him to take part in it. And once these assumptions are made, a scientist has to avoid in his own work anything that is not proper to the scientific way of thinking, and that means any reference to a God. C. F. von Weizsäcker writes (*World View of Physics*, p. 157): 'Even if the hypothesis of Laplace had been false in some particulars, still every scientist must certainly set himself the goal of making the hypothesis "God" superfluous in this field'. We could add to the word 'God' the whole concept of the numinous and all ethical considerations. Physics has been led to the Uncertainty Principle, not because the determinism of Newtonian mechanics left no space for free will, for right or wrong, or for a god other than the prime mover, or because such determinism was repellent and we needed an escape from it, but because the concept was forced on us by experimental observations on the behaviour of electrons made by Davisson and Germer in New York and by George Thomson in Aberdeen.

I said also that I thought science had contributed little to religion. At first sight this may be thought not to be so; there have been two centuries of conflict between science and what passes for religion and partly as a result the beliefs of most theologians differ today from those held by many of their predecessors at the beginning of this period. But other influences, notably the accessibility of the classical philosophers, have led in history to changes in the stories through which Christian theology is expressed, and the point that I would like to make is that neither philosophical authorities nor the progress of science can greatly change the essential content of religion. Both philosophy and science can embellish it; they can affect and beautify the stories (in the sense of Braithwaite's Eddington lecture[2]) and the images through which some men approach religious truth. I will have something to say about the contribution of science to this embellishment later. And of course science has done much to strip from religion some of the myths that have clustered round it. In the twentieth century we may laugh when we realise that the Aristotelian concept of the perfection of the supralunar world was once so closely associated with Christian belief that respectable people were shocked when in 1577 Tycho Brahe announced that the path of a comet, that intruder on celestial perfection, lay beyond the orbit of the moon and also when Galileo in 1610 claimed to have observed spots on the sun, as though they were blemishes on the face of the divine. We must remind ourselves that Aquinas and the great minds of his period craved for order, just as we do, and found it by this marriage of religion with the scientific knowledge of antiquity. Modern science has taught us that we must not seek it in this way.

I cannot find, then, that in the past science has helped religion in any positive way, nor religion science.

What of the future? How much the further advance of science will affect theology in time to come depends, it seems to me, on the status you ascribe to the stories of Christianity. Scarcely anyone now attributes to the Genesis story more than a symbolic significance in the light of modern geology and evolutionary theory. More worrying to Christian theologians is, I believe, modern research on the effect of deficiencies in diet on character, and its relation to original sin. Others feel that the doctrine of the incarnation would make it difficult to accept any future discovery of the presence of intelligent beings in other planets. If, however, one feels that the stories are aids to the understanding of the nature of God, 'to announce allegiance to a living God knowable through the stories of Jesus and the record of Christian experience', rather than events whose historical truth is important, these matters should trouble one little. None the less they trouble some people of high intelligence who have made a deep study of science, especially those who hold fast to the materialistic aspects of Christianity, according to William Temple the most materialistic of the great religions. Much as I disagreed with it I was impressed with a passage in a very recent book, E. L. Mascall's *Christian Theology and Natural Science*[3] in which he speaks of the role of space in theology. The New Testament writers, he says, quoting C. S. Lewis, managed to combine the idea of a heaven above their heads with the idea of the mode of life of an ever-blessed God without difficulty or confusion. With Newton's cosmology and his conception of Absolute Space, which, without relation to anything external, remains always the same and immovable, this was difficult; henceforth one had either to locate God and his angels in heaven in the same sense as Nelson's column is in Trafalgar Square, or else deny the legitimacy of applying spatial terms to them in any sense whatever. And he says that the consequences of the latter course are inevitable: 'Heaven becomes a purely spiritual state, the Ascension of Christ becomes a destruction of Christ's human nature, and our future condition becomes an enjoyment of immortality in which neither the body nor anything analogous to bodily experience can have any part.' And a further result, he says, is that complacent doctrine that 'heaven and hell are states, not places', which has done so much to eviscerate modern religion'.

From this dilemma the Einsteinian conception of space, Mascall says, has shown us a way out. Whether this is so or not, I can hardly say how strongly I disagree with Mascall's whole thesis. To do him justice Mascall elsewhere says quoting Dean Inge that in religion to marry the spirit of any age leads to widowhood in the next and quotes with approval the well-merited scorn of the late W. H. V. Reade on 'the näive delight sometimes exhibited by friends of religion when they hear that the solidity of matter is being whittled away by modern physics, as though Materialism would at any rate be less dangerous if only matter could be made decently thin'.[4] To my mind any introduction of spatial ideas into religion will lead to this form of widowhood, and if the absence of spatial concepts eviscerates religion, then perhaps we can put its guts back by some hard thought on its nature. If the leaders of religion are wise, it seems to me, they will formulate the doctrines of their churches in such a way that, though they provoke the intellect to the uttermost, the progress of science cannot touch their essence.

My thesis has been that science has added little to religion but has stripped away from it some concepts which, in our time and place, seem to belong elsewhere. My friend and fellow physicist Professor Charles Coulson, has written on this point in *Science and Christian Belief*,[5] and has warned that it is futile for religion to concern itself with those parts of knowledge, be it biology, psychology or cosmology, which science has not *yet* put in order. Of course I agree; here in Cambridge one need not labour this. But Coulson has something positive to say, something about a gift which science has actually made. He finds a synthesis of religion and science in the beauty and in the excitement of scientific discovery. 'The heavens declare the glory of God, the firmament showeth his handiwork', says the psalm. Coulson feels that the heavens declare it more gloriously when you can contemplate them with understanding of the birth of the spiral nebulae and the origin of the cosmic rays. Not everyone feels this; the cold machinery of Newton contrasted with the architectural universe of Aristotle and Aquinas is repellent to some. But it is a feeling which some of the greatest scientists have recognised. Einstein wrote that in his science he wanted to understand how God makes the world work, and that he was not concerned with little problems. John Ray, the founder of systematic botany and zoology, called his great book on these subjects *The Wisdom of God Manifested in the Works of the Creation*. He and Einstein would both perhaps have agreed with a sentence that I find in Canon Smethurst's book, *Modern Science and Christian Belief*,[1] that for the Christian the primary objective is the investigation of the wisdom of God manifested in the works of the Creation. It may well be that the contemplation of natural law can lead a scientist to religious feeling, since therein he may see the wisdom of God. If one wishes to maintain that the scientific approach to life is part of a broader religious approach, this is probably the best argument to use. But if Coulson or anyone else were to maintain that the day-to-day work of a scientist leads naturally to a religious approach to life, a desire to understand how God makes the world work, I would not agree. The trouble is that we are not most of us Einsteins, we are concerned with the little problems, not his great ones, and if we weren't science wouldn't progress. The little problems are those in which the consequences of the great advances are followed up, verified, and extended. In them we are concerned with the properties of the nucleus of the fluorine atom, the mechanism of creep in metal crystals or the synthesis of a vitamin. Worse, we are nearly all of us partly or exclusively concerned with the application of science. We are working for control over nature, we are trying to make things happen. And that has its dangers. May I quote you C. S. Lewis's diagnosis of our state of mind in his remarkable fairy story *That Hideous Strength*[6]: 'The Sciences, good and innocent in themselves, had begun to be subtly manoeuvred in a certain direction. Despair of objective truth had been increasingly insinuated into the scientists; indifference to it, and concentration upon power had been the result. Dreams of the far future destiny of man were dragging up from its shallow and unquiet grave the old dream of man as God.' The dream may be diabolical or not, that depends on your point of view. But the dream is there all right, in Russia and in America and here, the dream that we can control our destinies through the power of applied science. And even in our Cambridge laboratories, dedicated as some are to the pursuit of pure knowledge, we feel rightly that we must give society some return for the resources which it puts at our command. What is that return—the revelation of the wisdom of God manifested in the works of the Creation? He would be a bold man who made these the grounds of an application to the Department of Scientific and Industrial Research for funds to support his work.

Be that as it may, I do not believe that the day-to-day work of most scientists doing pure research has much to do with the appreciation of the values for which religion stands. I believe that when our aim is to apply it, the strategy of the application has even less, notwithstanding the aim of applied science to raise the standard of life. Although I agree that the contemplation of the greatest scientific generalisations may stand in a sense alongside religious experience, I would on the other hand even maintain that the day-to-day work of the scientist, the unravelling of one small detective story after another, the struggle with leaks, short circuits and mathematical tricks and the angling for financial support, gives him a training that is difficult to reconcile with religious thought. The trouble is that a scientist has such a definite idea of what he means by truth. Truth to him is what can be verified by experiment, what will serve as the basis of further experiments if he can get the money for them, and what is agreed by scientists on both sides of the Iron Curtain. Religious truth is so demonstrably not of this nature. It is a pity that the literature of religion is not written still in Latin; to have a different word for it, such as *veritas*, would help.

It seems to me that a scientific worker, conditioned by his profession as I have described, may approach religion in several ways. He will know that most of the great thinkers in the history of the world have been concerned in one way or another with the religious approach to reality. He may feel that in this age of scientific enlightenment this is a stage that we have outgrown. If so, it has happened from time to time that he has been enslaved by some other body of dogma, such as a belief in the perfectibility of man under Communist rule, which he will believe can unify science and all man's aspirations. Or, engrossed in his work and in his private life, he may just feel that there is nothing here for him, just as others of us may have no ear for music or no interest in sport.

But if, stimulated by his reading of history and literature or pushed on by some inner want, he turns to the religious aspects of things, what ways of entry are open to a scientist? He may prefer to study things and find his orientation outside an organised church. He may study the literature of ascetic theology. He may read, with the fascination that one feels for experience remote from one's own, of the Dark Night of the Soul, and learn with interest that in those religions which do not ascribe to the Deity the attribute of personality this path need not be trod. He will attempt to build his view of life on such wisdom as he can find in the literature of religion and conversation with his friends. Approaching the subject from the background of science he will not find his professional training useful in grasping these things. On the contrary, I have said that I believe that it stands in his way. But there is a compensation, that for a scientist this field of thought can have the fascination of the unknown and the unexplored. It is as though a man born and brought up in the suburbs of Cambridge should undertake a voyage to the strange landscapes of Japan, or Tibet, or Malacandra. Perhaps a man coming to religion from the foreign world of science may be struck more forcibly than another by its beauty, harmony and importance.

A scientist like anyone else, however, may wish to find what he needs within an organised church, within one of the churches of the Christian religion. I have said several times that I think that the day-to-day work of a scientist does not lead naturally to religious thought, and perhaps is a particularly bad introduction to the thought of the Christian churches. It gives him a criterion of truth which he will not find operative there. Both on the major plane, the existence of God and the divinity of Christ, and on the minor plane of the various miracles which are recorded in the New Testament and which are recalled in Divine Service, he will find the word truth used in a sense which is foreign to his way of thought,which is not susceptible to experimental test and which does not and cannot carry universal assent. He will find this even more in the ritual and creeds of the churches than in the literature of ascetic theology. What is he to do about this?

One way is to take a religion, be it evangelical, fundamentalist or Roman, and accept all that it teaches, mortifying the intellect thereby. Some scientists can do this. An outstanding and now laughable example was that of Sir Edmund Gosse's father, a devout man and a talented geologist, who recognised in the fossils the evidence for Darwin's theory of evolution but believed that the devil had put them there to try the faith of man. Few people would go as far as this now, but none the less one knows scientists whose outlook on life involves a dichotomy, who reject authority in science and believe that the scientific method will lead to truth, but abandon all this in religion and here believe that all truth, truth not *'veritas'*, truth in the everyday scientific sense of the twentieth century, is in the hands of their particular church.

I said at the beginning that, as a scientist speaking of religious things, I can speak with no scientific authority whatever. I can only give you for what they are worth one man's views on these matters. But speaking as one man, that is not the way that I would hope that my friends and pupils would go about it. I do not believe that we are required to mortify the intellect, to surrender any part of it. Let us start with the belief in God. The evidence of history, the evidence of literature, our own feelings in many cases lead us to believe that this is a meaningful concept to man. We think about it, we read about it and hear about it and a concept of God grows up in our minds. It cannot be expressed in scientific terms, it is perhaps different for each individual, it is difficult to express at all. The scientist is, I think, particularly ill qualified to explain it, because he is not used to using words to describe this sort of thing. In fact the scientist, I believe, when he wishes to contemplate God, must leave behind him the tools of his trade, his analytical method of thought. But as many men of scientific training have shown, it can be done without intellectual dishonesty.

But I am supposing that a scientist wishes to contemplate and worship God in company with other like-minded persons, within the walls of a Christian church. Here, in the rites of the church, he will find many statements which go beyond the existence of God and the expression of his nature in the life of Jesus Christ. He will be asked, when he says the Apostles' Creed, to assent to several statements in addition to this. Can he do this and guard his intellectual integrity?

I would say at once that one cannot on strictly scientific grounds object to the belief that, at the birth or death of a divine person, miraculous events occurred. The event was, in the belief of Christians, unique; science deals with events that can be repeated; so science cannot properly say anything about them. My own strong disinclination to accept these miracles is, I believe, based on aesthetic rather than scientific grounds; it springs from the feeling that they are unnecessary. The life of a great man is such a miracle in any case that I cannot conceive that the Ground behind all being would mark it with the kinds of miracle that are related.

Nevertheless I would not wish any substantial change made in the historic Creed and form of service of any church which I attended; I would like to conclude my talk by explaining why this it so. 'Faith', said some anonymous schoolboy, 'is believing what you know ain't true', and may I describe one approach which a scientist may take to parts of organised worship as 'loving what you know ain't true'. When we hear the sonorous words of the creed describing great miracles, we may believe them as scientific truth, or as another kind of truth which lies outside science, or, to use Braithwaite's phrase again, as stories which to a logical empiricist signify commitment to a charitable way of life. Or maybe they remind us comfortably of school chapel and past experiences of that kind. But if they appear to any scientist an affront to his sense of what is beautiful and right, then to him they will mean none of these things. I suggest that even then they can have one important significance: they are part of the history of religion. Dr Thornton in his book *Revelation and the Modern World*[7] has described how the theological movement known as liberalism broke down, through its fundamental assumption that the interior content of the Christian revelation could be detached without distortion or mutilation from the outward form in which it was originally given and could be inserted, again without distortion or mutilation, into the thought-forms of the modern age. Liberalism, says Dr Thornton, 'set out to preserve the unchanging essence of the Biblical revelation; but it too easily, and even naïvely, identified that essence with the contemporary philosophic idealism of Western man. In doing so, it evaded the whole problem of historical religion, the problem as to how eternal truth can be manifested at all in the infinitely complex, slow-moving yet ever-changing processes of time and space'. I believe Thornton to be right on this point; one can only know religion through the records of

Christian and other religious experience; history and what has been written throughout time is of the essence of the matter. I believe that the function of a church and of Divine Service is to bind together not only the worshippers at one time and place but the worshippers at all places and at all times from the past up till the present. I believe that these old forms of words do this—and that a scientist approaching the church in the way that I have described can take them in this sense, as a link with a long history of religious endeavour.

[1] A.F. Smethurst, *Modern Science and Christian beliefs* (Nisbet, 1955).

R.B. Braithwaite, *An Empiricist's View of the Nature of Religious Belief* (Cambridge, 1955).

[3] E.L. Mascall. *Christian Theology and Natural Science*, (Longman, 1956).

[4] W.H.V. Reade, *The Christian Challenge to Philosophy*, SPCK 1951.

[5] C.A. Coulson, *Science and Christian Belief* (Oxford, 1955).

[6] C.S. Lewis, *That Hideous Strength* (John Lane, 1945).

[7] L.S. Thornton, *Revelation and the Modern World* (Dacre Press, 1950).

Rocking the Cradle of Britain's Nobel prize babies

From *The Economist*, 27 Feb. 1982, with permission.

Britain has performed consistently well in science partly because it has nurtured centres of excellence. One of those centres—the Cavendish Laboratory in Cambridge—is now facing a difficult future, perhaps one with fewer Nobel prizes.

Science is one of the few activities in which Britain has excelled during the past generation. Measured by Nobel prizes or (more prosaically) by papers published in leading journals, Britain has been second only to the much larger United States. That position may now be slipping. Britain's share of papers in influential journals dropped during the 1970s.

Britain's most outstanding research lab is the Cavendish Laboratory in Cambridge. Cavendish scientists were involved in about 25 Nobel prizes over the past 80 years: as many as all scientists in France. They won four Nobel prizes in the 1970s alone.

The lab specialises in pure, rather than applied, physics. To some, this excellence at pure science may seem symptomatic of the distorted, anti-commercial priorities that have led to Britain's decline. It is not. The frontiers of science being explored at the Cavendish will help lead to new microelectronic devices, new chemical catalysts and new engineering principles. It is symptomatic that American industry seems more interested than British industry in the work being done at the lab.

The next four pages explore the Cavendish's success and ask whether it can continue. One conclusion is that, even when science budgets were expanding, the laboratory was ruthlessly pruning good research to make way for even better research. The lab seems at present to have lost some of the will needed to go on doing this.

There is a lesson here for British universities generally. During the expansionist 1950s and 1960s, they had room for everything, the good and the bad. During the 1970s, stagnation hit them. They muddled through, failing to reorder their priorities, to create openings for younger scientists and to replace obsolete equipment. There is a backlog of necessary pruning. Budget cuts are now forcing savings to be made. Few universities are cutting more than they absolutely have to, in order to make the books balance. The backlog of desirable pruning is being ignored.

The Cambridge climate

The Cavendish is in modern buildings on the outskirts of Cambridge. The focal point is the common room, a spacious cafeteria where everybody gathers without fail for coffee at 11:00 and tea at 4:00. Each research group has its regular table. There are blackboards, in case anybody wants to write equations or draw diagrams. The atmosphere is relaxed and open but underneath there is an intense competitiveness characteristic of Cambridge science. The coffee and tea breaks are soon over.

Meet Dr Roy Willis, a 39-year old physicist on whom the Cavendish is pinning high hopes. He says of the Cambridge climate:

> This university worships success. It uses the people most likely to gain success ... You fail here if you don't get the Nobel prize. It is always self-analytical. Somebody once said that, joining a new school in America, you have to learn to use your fists; here, you have to learn to be kicked intellectually.

Dr Willis cut his research teeth at the Cavendish in the 1960s and then joined the European space programme. A priority there was to find out how materials would behave in space and this led to new techniques for studying atoms on the surfaces of materials. Surprisingly, surface atoms turned out to behave quite differently to those in the bulk of a material. Studying them became a hot area of physics and Dr Willis became one of the hottest people in it.

In 1979, Royal Society money enabled the Cavendish to attract him back, at a large cut in pay, at a relatively junior level (he was being offered professorships elsewhere) and with nothing more than some laboratory space and a technician. He came because Cambridge offered freedom to think and a challenge.

So far, Dr Willis has not earned that Nobel prize. He thinks, if he does not do something exciting at Cambridge, he might resign rather than rest on the laurels of a tenured position. In his field, he reckons, he is competing with about 60 scientists worldwide, any one of whom might pull an ace out of the pack.

As well as doing research, the Cavendish is the university's physics department. It has to teach nearly 800 undergraduates. A quarter will go on to do a PhD, a quarter will do other courses for a year and a quarter will go straight into industry. Half the PhDs may eventually end up in industry.

The dual role of teaching and research is reflected in the laboratory's budget. Some £2.2m comes from the University Grants Committee (UGC), the body which parcels out finance for university teaching in Britain. Another £1.6m is research grants, mostly from the Science and Engineering Research Council (SERC), which allots the money for basic physics research in Britain.

The dual role of the Cavendish contributes to its intellectual climate. Teaching forces a researcher to clarify his ideas; keeping ahead of the brighter undergraduates is a challenge. There is a price. One problem is that there are things the UGC and the research councils both think the other should pay for—so for which neither pays.

Also, the Cavendish scientists are competing with researchers elsewhere who do not have teaching duties. In other countries, there are institutes devoted entirely to research at the frontiers of physics. Britain has nothing comparable to the Max Planck Institutes in West Germany or Bell Laboratories in America.

The international nature of modern physics requires that scientist be away from base a lot of the time. This, too, imposes a strain. Cambridge University insists that faculty members reside in Cambridge during term time. Five Cavendish faculty members are currently listed as on leave of absence.

Altogether, there are about 50 faculty members at the Cavendish, plus over 200 research students and post-doctorates. All but seven of the faculty have a Cambridge degree. This incestuousness is understandable. The university has access to the cream of science students in Britain. The scientifically ambitious are drawn to Cambridge, just as the politically ambitious are drawn to Oxford. Some 30% of all the fellows of the Royal Society are Cambridge alumni, three times as many as from Oxford (though Oxford has about the same number of science students).

The growth of new universities in Britain has not narrowed Cambridge's lead; rather the reverse. Among the younger fellows of the Royal Society, half were educated at Cambridge.

Cambridge tends to attribute its success not to the Cambridge soil but to the quality of the intellectual seeds planted there. Sir Sam Edwards (acting head of Cavendish) says one must ask not why Cambridge does so well in science but whether it does well enough.

This typically self-critical remark—Cambridge is a mixture of self-criticism and intellectual arrogance—says something about the Cambridge climate. It is a hot-house. For example, professors compete with lecturers for research students. Research students vote with their feet. At Cambridge, a lecturer can wind up with more research students than a professor.

To be sure, Cambridge has its share of dead wood. The ancient colleges are bastions of tradition. They do much dining but play a secondary academic role. Departments, not colleges, choose who gets a tenured job and this is a competitive business. After getting his doctorate, a scientist will spend five years as a research fellow. Then he may be offered a trial period as a teacher. This, too, lasts five years and the rejection rate at the end is high. Only when he has passed this test will he get tenure, usually in his mid-30s when he has proven not only that he has promise but also that he can fulfil it.

Cambridge pays more than lip service to the idea that a lecturer should be good at teaching, but there is a let-out enabling some bright people who are bad at teaching, or just disinterested, to be retained. There are some tenured research posts that require minimal teaching.

Running the lab

Providing opportunities for people like Dr Willis requires leadership. When Lord Rutherford ran the Cavendish from 1919 to 1937, he was a dictator. People quaked when he walked round the lab. Things have changed. Rightly and inevitably. For one thing, physics does not have men of Rutherford's stature any more. The honours are more equally shared.

Perhaps the nearest approximation to Rutherford is Sir Nevill Mott, head of the Cavendish from 1954 to 1971. Now aged 77, he is still doing research. A sharp-minded eminence grise, he is taciturn but what he says cuts.

Sir Nevill was blooded at the Cavendish in the 1930s, growing up during the explosion of the quantum revolution, aptly known as the golden age of physics. A colleague says, meaning it disparagingly, that Sir Nevill thinks physics should always be like that. Perhaps, though, some such belief is needed to inspire the greatest physics.

After Cambridge, Sir Nevill moved to Bristol where he did some of the key work that led to the invention of the transistor. He was unlucky not to get a Nobel prize for that.

When he came back to run the Cavendish, he made it virtually a full-time job. In his 60s, though, he decided to have a crack at an unexplored subject: non-crystalline materials like glass. Compared with perfect crystals, such materials seemed an impenetrable mess. Sir Nevill showed they were not messy after all and started a new branch of physics with potential applications in solar energy. This did get a Nobel prize. Few scientists win this accolade for work in their 60s.

Sir Nevill's successor as head of the Cavendish describes the job as being to encourage the good when you see it. There is another side, says Sir Nevill. To encourage the good, you have to close down the beta plus. 'That's the hard part of the job.'

It is one Sir Nevill did not shrink from. Before he had arrived back in Cambridge a decision had to be taken about an accelerator being constructed to study fundamental particles of matter. Distinguished scientists at the Cavendish were carrying on Rutherford's tradition in this. Sir Nevill felt particle accelerators would soon get beyond Cambridge's pocket. He stopped the accelerator.

He then insisted that molecular biology, a subject the Cavendish had pioneered with the discovery of the double helix, should find another nest. He felt its growth would be like that of a cuckoo, and crowd out physics. He also decided to run down crystallography, a physics branch founded by the Cavendish. It takes guts to close such things.

In their place, he built up two subjects that had been started by his immediate predecessor; radio astronomy and solid state physics. It is in these subjects that the Cavendish had its four Nobel prizewinners in the 1970s.

Sir Nevill built up muscle in subjects by pinching good research groups from elsewhere. 'It was like running a company', he says. 'When I saw an opportunity for a good takeover, I took it.' He believes the laboratory needs a 'full-time head who thinks his place in heaven depends on making the right choices for the laboratory'. This is what the department lacks at present.

One current choice that Sir Nevill dislikes concerns a protégé of his: Dr Michael Pepper. Dr Pepper arrived at Cambridge in 1973, on sabbatical from Plessey, the British electronics firm. Like Dr Willis, he is studying the surface of materials and, though a less forceful personality, he is another hot property in this field. He uses a technique that enables you to stack very thin layers on a surface and so give a material new properties.

Industry is interested. Fujitsu thinks it can develop a rival to Josephson junctions for the ultra-fast circuits needed in the next generation of computers. Josephson circuits—another Cavendish idea—must be cooled to around minus 270°C. IBM is pouring money into them. Fujitsu reckons its circuits will be just as fast at higher temperatures.

The technique behind this, and behind Dr Pepper's work, is called molecular-beam epitaxy (MBE). A good MBE machine costs £400,000—more than Cambridge felt it could afford. With some reason. That figure is nearly twice the lab's annual equipment grant from the UGC.

Dr Pepper has access from time to time to a machine run by SERC for universities' joint use, but the SERC machine is under-staffed and getting time on it is a bureaucratic hassle. Dr Pepper fears he is going to fall farther and farther behind the international competition. Besides, the Cavendish has not yet shown any sign of offering him a tenured post.

Rightly or wrongly, Sir Nevill would have backed Dr Pepper, cutting back elsewhere if necessary. And Dr Pepper's case points up a problem that will recur. The time has come when a series of tough decisions look like being needed. Apart from the gloomy outlook for university finances, the subjects that Sir Nevill built up—partly because they were cheap at the time—are becoming very expensive.

A new style

The tone of the Cavendish since Sir Nevill's time was set by his immediate successor, Sir Brian Pippard, who stepped down only recently and is still very much a figure at the lab. He brought a change in both style and philosophy.

In Sir Brian's view, very few important decisions are taken by the department head. 'This department', he says, 'is a headless department that runs merely by custom: strong custom'. He took the appointment for a limited period and concentrated on reorganising teaching rather than research.

He did a good job on the teaching. He believes fervently that teaching is the lab's first duty, that it owes this to the brilliant students who entrust Cambridge with their education.

When Sir Brian first joined the Cavendish in the 1940s, the great men there seemed to him—for all their talents—gentlemen amateurs. He sympathised with those trying to make the laboratory more professional. He now feels the process has actually gone too far.

In the 1940s, a scientist could make his own equipment. Modern experiments often need an army of researchers who can compete only if they have the biggest machine. (By contrast, Sir Brian has only once had a research grant during a successful research career.) There is cheap stuff you can still do in astronomy but it requires hard thinking: spending money is easy.

Unfortunately, the days of string-and-sealing wax physics are past. There is a choice between a £0.4m machine and a £4m machine but without the £0.4m machine you may as well give up. Without it, you would fulfil a fear that Sir Brian has—that physics is getting to the stage where there will not be exciting enough problems to attract the young; that it will become a dull subject like heavy electrical engineering.

Sir Brian is well aware that the time has come for the Cavendish to rethink its priorities again, but he favours devolving responsibility for this. He is telling the people in his own group to think the priorities out for themselves. A dictator, he says, would cause enormous anger.

A dictator is one thing. A strong leader might be welcome in the laboratory. In 1979, Sir Brian handed over to Professor Alan Cook as acting head of the Cavendish. Professor Cook is talked of as a good committee chairman. Whatever his qualities, they are not now in Cambridge. He has gone to California for a 10-month sabbatical, just when Cambridge is in the middle of a university crisis.

Left keeping Professor Cook's job warm is Sir Sam Edwards. Sir Sam has the scientific imagination and the courage to lead. He is an ex-chairman of SERC, where he sometimes went against the current but was proved right more often than not. He recently chaired a committee that advised London University to close five or six of its 10 physics departments. Whether they will take the advice remains to be seen.

Unlike Sir Brian, he thinks the head of the Cavendish still is powerful. He reckons that Cambridge's committee structure puts power into the hands of somebody who really wants to wield it. However, as only acting head of the Cavendish, Sir Sam reckons he cannot take strategic decisions.

The lab does have an able manager, Mr John Deakin, its secretary. He is an engineer with industrial experience. He takes on most of the administrative load that normally falls to heads of departments, leaving professors free to get on with teaching and research. It is an excellent arrangement. But there are some decisions only a department head can take.

References

1. Gamow, G., 1970, *My World Line: An Informal Autobiography* (New York: Viking Press).
2. Mott, N. F., 1930, *An Outline of Wave Mechanics* (Cambridge University Press).
3. Mott, N. F., 1980, The beginnings of solid state physics. *Proceedings of the Royal Society*, A371, 1.
4. Jones, H., Mott, N. F. and Skinner, H. W. B., 1934, *Physical Review*, 45, 379.
5. Born, M., 1978, *My Life: Recollections of a Nobel Laureate* (London: Taylor & Francis).
6. Badash, L., 1985, *Kapitza, Rutherford and the Kremlin* (Yale University Press).
7. Jones, R. V., 1978, *Most Secret War, British Scientific Intelligence 1939–45* (London: Hodder and Stoughton).
8. Casimir, H. G. B., 1983, *Haphazard Reality — a half century of science*, (New York: Harper and Row) (autobiography).
9. Gowing, M., 1974, *Independence and Deterrence; Britain and Atomic Energy 1945–52* (London: Macmillan), p.84.
10. Ashby, E., *Technology and the Academics*.
11. Powell, F. C., 1984, *The Caian*, Nov. 1984.
12. Bloch, F., 1928, *Zeitschrift für Physik*, 52, 555.
13. Wilson, A. H., 1931, *Proceedings of the Royal Society*, A133, 458.
14. Brinkman, W. F. and Rice, T. M., 1973, *Physical Review B* 7, 1508.
15. Mott, N. F. and Twose, W. D., 1961, *Advances in Physics*, 10, 107.
16. Mott, N. F., 1969, *Contemporary Physics*, 10, 125.
17. Adler, D., Henisch, H. K. and Mott, N. F., 1978, *Reviews of Modern Physics*, 50, 207.
18. Brock, W. H. and Meadows, A. J., 1984, *The Lamp of Learning* (Taylor & Francis).
19. Stockwood, M., 1982, *Chanctonbury Ring* (London: Hodder and Stoughton).
20. Popper, K., The Open Universe—an Argument for Indeterminism in *Postscript to the Logic of Scientific Discovery*, edited by W. W. Bartley, Volume III (London: Hutchinson and Co.), pp. XXII–185.

Index